DATA
BROADCASTING
Revised Edition

DATA
BROADCASTING
Merging Digital Broadcasting with the Internet
REVISED EDITION

Lars Tvede
The Fantastic Corporation, Zug, Switzerland

Peter Pircher
The Fantastic Corporation, San Francisco, USA

Jens Bodenkamp
ETF Group Deutschland GmbH, München, Germany

JOHN WILEY & SONS, LTD
Chichester • New York • Weinheim • Brisbane • Singapore • Toronto

First edition published in 1999 as Data Broadcasting – The Technology and the Business, by John Wiley & Sons, Ltd.

Copyright ©2001 by John Wiley & Sons, Ltd,
Baffins Lane, Chichester,
West Sussex PO19 1UD, England
National 01243 779777
International (+44) 1243 779777
e-mail (for orders and customer service enquiries): cs-books@wiley.co.uk
Visit our Home Page on http://www.wiley.co.uk or http://www.wiley.com

Other Wiley Editorial Offices

John Wiley & Sons, Inc., 605 Third Avenue,
New York, NY 10158-0012, USA

WILEY-VCH Verlag GmbH
Pappelallee 3, D-69469 Weinheim, Germany

John Wiley & Sons Australia, Ltd, 33 Park Road, Milton,
Queensland 4064, Australia

John Wiley & Sons (Canada) Ltd, 22 Worcester Road,
Rexdale, Ontario, M9W 1L1, Canada

John Wiley & Sons (Asia) Pte Ltd, 2 Clementi Loop #02-01,
Jin Xing Distripark, Singapore 129809

British Library Cataloguing in Publication Data
A catalogue record for this book is available from the British Library
ISBN 0471 48560 8

Typeset in Garamond by Deerpark Publishing Services Ltd., Shannon
Printed and bound in Great Britain by Biddles Ltd., Guildford, Surrey

This book is printed on acid-free paper responsibly manufactured from sustainable forestry, in which at least two trees are planted for each one used for paper production.

Contents

Foreword

The combination of powerful computing devices, high-speed data communications and a new class of interactive applications is creating the first new medium since television, which was born over 50 years ago. In many ways this new medium will encompass the capabilities of television and print and will combine them with the interactive and targeted capabilities of the telephone. It will impact on all forms of human communications including entertainment, education, commerce, information and personal communication.

One of the key enabling technologies (really a very large set of technologies) is the Internet. The Internet is at a very early stage of its development and in many ways is still very unpredictable and surprising.

One of the reasons it has developed so fast is that it is extremely flexible and adaptable. It has been able to utilise infrastructures, content and technologies that were developed for other purposes in a truly extraordinary way. For instance, the key infrastructure that it uses as a transport vehicle (to move information from one place to another) is the telephony system, which was designed for a totally different purpose (point to point voice communication). With the exception of a small number of homes that are served by cable modems or via satellite, consumers are connected to the Internet via the telephone system. As a result, the telephony systems' evolution is now driven in a large part by the needs of the ever-expanding Internet.

Many of the limitations of the telephony system have hindered the power of this new interactive medium. Most consumers at home can utilise the Internet only at data rates that are extremely low (56 kbps and under) and not well matched for the devices that they use (mostly PCs at this point) and the requirements of the applications. Not only are they mismatched, but the gap is growing. PCs and other devices improve their performance based on Moore's law (performance doubles every 18 months) while communication speeds barely increase by four times in 5 years. Technologies like ADSL promise to improve the performance of the telephony network by as much as 100 times.

Cable modems also hold great promise to provide similar types of bandwidth using an infrastructure that was created for an entirely different purpose, television. Countries like the United States, Canada, and the Netherlands will benefit from the ubiquity of both the cable and telephony networks. But when we consider the future of bandwidth and how it will impact on the development of media, we have to realise that different parts of the world are in very different states of infrastructure development. Some parts of the world have both cable and telephony. Other parts, like most of Europe, are served by telephony networks and satellite. And while most of the people in the world have access to TV (TV home penetration in China is 90% in urban areas, for instance), less than half the world's population has ever made a phone call and only about 40% of the world's homes have a telephone. The bulk of the world will be brought into the digital age not through telephony but through their television systems. The bulk of television content will be delivered via satellite. There is a huge amount of bandwidth in most television broadcast systems. Cable and satellite systems can deliver gigabits of information. However, since this bandwidth must be shared, it is most efficient to broadcast information that many consumers would like at the same time. This information can then be stored locally and used when required.

Since 1991 I have led Intel's activities in furthering the development of residential broadband communication. In the process, I have had the opportunity to work with many of the pioneers in this field. Three such individuals have written the book you are about to read. They are pioneers indeed and have the arrows in their backs to prove it. Dr Jens Bodenkamp, an Intel employee, worked closely with me in Europe as we explored every avenue available to provide high-speed communications to homes. Along the way, we met Lars Tvede and Peter Pircher, who were co-managing a company focused on developing the tools that content owners would need to function in this new medium. They called their company The Fantastic Corporation, a demonstration no doubt of their enthusiasm.

We are about to enter the broadband world. It will require new applications, development tools, devices and communication infrastructures. It holds the promise to change all of our worlds no matter where we live and what we do. It will impact on the way we learn, on the way we play and the way we communicate. But a lot of hard work will be required along the way

This book will take the reader into the world of broadband and its implications for the future. Enjoy the trip.

Avram Miller
Vice President
Intel Corp.
1999

Acknowledgements

In many ways, this book was a collaborative project. It depended on the willingness of many people to provide us with information and comments. We would like to extend our profound thanks to all of the individuals who generously provided their assistance in reviewing this book, and in particular to Eva Parilla, Wen Liao and Tamara Grahn, who spent nights and weekends going over some of the sections. We also wish to thank the following individuals for their assistance:

Frank Ewald
Eric Troelsen
Heidi Zürcher-Krieger
Tony von Rickenbach
Elena Jeung-Branet
Cristina Vanza
Manish Bhatia
Gerard Wisemann
Henrik Schonau Fog
Leng Stricker Wong
Sara Watkins
Torbjörn Winther

Your contributions made a difference!

Lars Tvede
Peter Pircher
Jens Bodenkamp
Zug, Switzerland, April, 1999

List of Tables

List of Figures

1. The Evolution of Broadcasting

"Professor Goddard does not know the relation between action and reaction and the need to have something better than a vacuum against which to react. He seems to lack the basic knowledge ladled out daily in high schools."

1921 New York Times editorial about Robert Goddard's
revolutionary rocket work

This book is about a new business and technology called ''data broadcasting''. Data broadcasting is a concept whereby:

- A combination of video, audio, software programs, streaming data, or other digital/multimedia content...

- ...is transmitted continuously to intelligent devices such as PCs, digital set top boxes and hand-held devices...

- ...where it can be manipulated.

The broadcast concept means that although a return path might be available, it is not required. Content is received without being requested.

Data broadcasting is extremely user friendly and is suitable both for the work/study situation (''2-ft media''), when we travel (''mobile media'') and for the living room (''10-ft media''). In bringing broadcasting and the Internet together data broadcasting provides web content, live video, surround sound music, etc. – combined together, all on one device.

There are many services and applications that can be operated within a data broadcasting system. It can be used for background routing of large e-mails, fast delivery of content to Internet Service Providers and for a new and very powerful way of videoconferencing in corporations. Another data broadcasting concept is that of a "channel": a constantly updating media experience that combines rich broadcast with interaction. Such a "channel" need not be confined to single media; it can have streams of video in one part of a screen and interactive content of any sort in other parts, and can provide the ultimate electronic entertainment. Data broadcasting can also provide a range of other experiences, like virtual universities, powerful corporate presentations, etc.

Data broadcasting did not just come out of the blue – it is the natural next step in the evolution of broadcasting as this world moves from analogue to digital. To understand why this is so, it is perhaps useful to go back a bit in time and to get the context of how broadcast media actually developed.

1.1 THE CONCEPTION OF ELECTRONIC BROADCAST MEDIA

The world has seen a number of successful new electronic media being born over time, and each of these new media have tended to go through a number of rather similar phases. A new medium, begins as innovators demonstrate the ability to get the basic technology to work.

> *"There is a hole through! It's transmitting!*
> *It's the beginning of a new era!"*

This is the *technical* beginning. But then, perhaps to the innovators' surprise, follows a period when nothing much happens. Nothing until someone finally raises sufficient funds to launch the related applications and services commercially. That is the *commercial* beginning.

While the first media experience the technology offers represents a technical breakthrough, it is often not very exciting from a user perspective, as the tendency in the initial delivery is largely, well, just more of the same – the delivery of content made for previous media, now just delivered in a new way. For instance, when the radio was invented it was initially largely used for reading books aloud. The first content for television was often theatre pieces. The Internet was first used to transmit plain text in the form of e-mails.

This early technical phase represents an experimentation with the new technology until the commercial models fall into place and talented media people join the scene and implement the real opportunities of the new technology. Only then can this medium take off. This is the beginning of a new *medium*, and with this the real business.

1.1.1 Something in the Air

The first electronic medium to reach mainstream was the radio. The man who initiated it was an Italian by the name of Guglielmo Marconi (1874–1937), who in 1899 transmitted the first radio signal across the English Channel to France.

In the early 1920s, the British government provided licences to some 6,000 specialists experimenting with radio. These early adopters were playing with the new equipment not because of the programming available (there wasn't any), but because they were fascinated by the technology. The next phase began in 1922 when several manufacturers of radio equipment approached the British government suggesting that they should found a broadcasting company. By then the technology had been thoroughly tested, it worked well, and it was time to utilise it for the benefit of the general public. The six largest manufacturers co-operated to set up the British Broadcasting Company (BBC) which started transmitting programmes regularly in November 1922.

The venture was a huge success, and it was also an early indication of the potential commercial power of electronic broadcasting concepts. Within a year, several hundred thousand listeners were tuning into the BBC. Surprised at the overwhelming success of this new medium the government bought out the initial shareholders, and it was not long before the BBC had established itself as a leading global broadcasting organisation.

Meanwhile, in a commercial environment very different from the British, radio broadcasting was spreading fast and sure throughout the USA. The United States, having minimal regulation, created a very diverse and dynamic environment for radio to expand and evolve within. Radio became a patchwork of channels broadcasting entertainment of a very wide variety – from the best to the worst.

1.1.2 The Beginning of Television

The next big wave was television. In 1926, a Scotsman, John Logie Baird, introduced the world's first ''televisor'', and the British were again first to implement the medium. In 1929, the BBC sent out the first test signals on medium wavelength, and in 1936 they set up the world's first television service. In the years that followed most governments in the world launched national television channels. However, the sales of television sets did not take off until after World War II.

In 1970 the number of television sets had reached almost 300 million, and it kept growing. Until the 1980s television was largely confined to the OECD countries and the Soviet Union, but since then, penetration in other areas of the world (mainly China) has been increasing rapidly. The growth within the last few years was partly driven by the introduction of satellite technology.

1.1.3 The Age of Satellites

The story of satellites for television starts quite a few years before its first technical implementation. Back in the 1940s, a science fiction writer had already come up with a very interesting idea. In an article published in the magazine *Wireless World* ("Extra Terrestrial Relays"), Arthur C. Clarke stated what many people knew:

> *"A rocket which achieved a sufficiently great speed in flight outside the earth's atmosphere would never return. This 'orbital' velocity is 8 km/s (5 miles/s), and a rocket which attained it would become an artificial satellite, circling the world forever with no expenditure of power – a second moon, in fact."*

So you could send a satellite into orbit, and it would automatically stay up there, simply circling around. Clarke continued from this basic observation:

> *"There are an infinite number of possible stable orbits, circular and elliptical, in which a rocket would remain if the initial conditions were correct. The velocity of 8 km/s applies only to the closest possible orbit, one just outside the atmosphere, and the period of revolution would be about 90 min. As the radius of the orbit increases the velocity decreases, since gravity is diminishing and less centrifugal force is needed to balance it."*

This was still all standard physics. But now he described his new idea:

> *"It will be observed that one orbit, with a radius of 42,000 km, has a*

period of exactly 24 h. A body in such an orbit, if its plane coincided with that of the earth's equator, would revolve with the earth and would thus be stationary above the same spot on the planet. It would remain fixed in the sky of a whole hemisphere and unlike all other heavenly bodies would neither rise nor set. A body in a smaller orbit would revolve more quickly than the earth and so would rise in the west, as indeed happens with the inner moon of Mars. Using material ferried up by rockets, it would be possible to construct a 'space-station' in such an orbit.''

But what would such a space station be used for? Here is what he proposed:

"Let us now suppose that such a station were built in this orbit. It could be provided with receiving and transmitting equipment (the problem of power will be discussed later) and could act as a repeater to relay transmissions between any two points on the hemisphere beneath, using any frequency which will penetrate the ionosphere... A single station could only provide coverage to half the globe, and for a world service three would be required, though more could be readily utilised.''

So what Clarke explained was how his space station would act as an antenna for telecommunication to satellite dishes, thus overcoming the traditional problem connected with Earth-based antennas: the curvature of the Earth.

Clarke's vision was only science fiction. But then in 1957 Americans were shocked to learn that the Soviet Union had launched the first Sputnik rocket. This was the beginning of the satellite race, as the USA and the Soviet Union each invested enormous resources in research, both fearing the other party would gain a decisive military lead.

In 1962, NASA sent the world's first communications satellite into orbit. As the satellite moved above the surface of the Earth, it transmitted pictures from NASA's control centre in Maine, a greeting from the UK, and live pictures of Yves Montand singing in France. However, each control station was able to connect to the control centre only for a short moment as the satellite moved in and out of range.

Clarke's calculations of the correct orbit for geostationary satellites were not exactly correct (the correct distance is approximately 36,000 km), and his estimate for the dish sizes needed for receivers (''1 ft'') were opti-

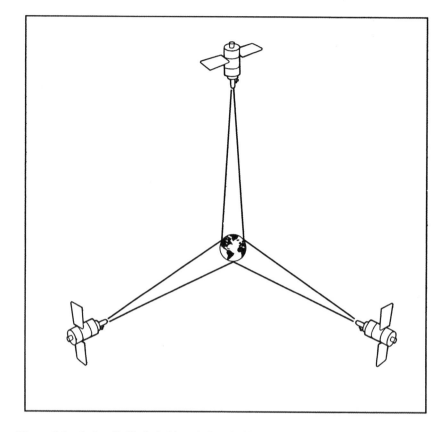

Figure 1.1   Arthur C. Clarke's idea. Arthur C. Clarke was the first person to understand the potential of using geostationary satellites for telecommunication

mistic when compared to what was initially possible (5 m). But the vision was correct. As history has shown the market developed rapidly, and probably much faster than he had imagined. One of the prerequisites for this dynamic new market was the founding by a number of governments of the Intelsat consortium in 1964 (Figure 1.1).

The purpose of Intelsat was to co-ordinate the development of international satellite networks. One of the tasks of the organisation was to allocate space in what is now known as the "Clarke Ring"; the "geostationary" path 36,000 km above the Equator. Intelsat's satellite services ranged from telephone calls to data transmission. By the 1970s more than 100 nations had joined Intelsat, and soon over 70% of the world's international telecommunications passed through Intelsat's network. Another major satellite consortium was Eutelsat, which was owned by the European PTTs.

These two consortia initially dominated the entire satellite markets. They represented governments and public institutions and it was therefore important to them that satellites did not compete with terrestrial telephone lines. Consequently, Intelsat's and Eutelsat's satellites were not intended or designed for people to receive data direct into their homes (which would have competed with local telco lines). The first satellites were primarily used either for military purposes or for transmitting long-distance telephone conversations between different telco networks. In the mid-1970s, receiver stations for receiving signals from Intelsat satellites, i.e. "down-links", had to be 5 m in diameter.

1.1.3.1 Satellite Operators

Gradually with the deregulation of the telecom industry, the geostationary satellites found use for new commercial applications, and private companies were able to obtain access to capacity ("transponders") through "operators". Operators were essentially middlemen who bought satellite capacity and resold or used it for different purposes. Negotiating with the PTT-owned satellite operators tended to be difficult, partly due to bureaucracy within the complex ownership structure, and also because the PTTs wished to maintain their monopoly position with respect to terrestrial communications. The most difficult negotiations were probably those with Intelsat, as they involved all the countries in the consortium. PanAmSat, on the other hand, was an early example of the reverse situation: a company thinking along commercial lines, and ready to move quickly. PanAmSat was launched in 1988, after years of court battles and disputes.

Another obstacle in the satellite market was the need to get permission to install hardware (satellite dish) at each individual end-user. In Europe the European Community removed that barrier in 1990. However, it still exists in some parts of the world even though the global trend is towards increased liberalisation.

A final hurdle was related to the broadcasting of data. In order to get broadcast permissions the new satellite companies had to accept some limitations. A typical restriction was that they had to agree not to offer data services, essentially so that the telecoms market would remain protected from competition.

However, with advances in technology and mounting pressure from governments and private companies restrictions on data transmission have rapidly been reduced. An important change took place in the USA under the Nixon administration. A commission headed by Clay Whitehead was set up to design a new policy for the utilisation of satellite communications. The result of the commission's work was the so-called "Open Skies" policy allowing privately owned businesses to obtain permission to launch satellites for private purposes, which, of course, they immediately started doing. Soon the Clarke Ring was full of satellites supplying everything from telephone calls to home banking to TV programmes. However, the signals were still not direct-to-home ("DTH"), as the average satellite dish required to receive signals was over 2 m in diameter. Yet in 1976 a satellite was launched for transmission of TV direct to private homes. It was Canadian and its name was "Hermes".

In 1982, 6 years later, the same process started in Europe, and in December 1988 a very important European TV satellite, Astra 1A, was put into orbit. A private consortium, "SES", owned Astra. The up-link was established in Luxembourg where legislation was very liberal. To acquire the right to up-link, SES had to convince Intelsat and Eutelsat that Astra would cause no "technical and economic harm" to them.

1.1.3.2 Typical Private and Public Satellite Services

With a greater number of satellites and with falling unit prices, the satellite business started to move forward very quickly and a commercial market structure soon evolved, as satellites were able to perform a wider range of commercial services. The most important ones of these were:

- *Broadcasting of data and video*. Broadcasting of data and video was effected as point-to-multipoint communication, where either the receiver station could be large dishes in "hubs" (data centres) from where the signal was re-transmitted, or the signal could be broadcast direct-to-home or via cable.

- *Switched data systems*. Switched data systems were networks of interactive Earth stations, all of which were able to communicate with any one of the other stations. The name "switch" makes one think of a switchboard, which was basically the underlying concept (the signal could be freely directed from one recipient to another). Switched data systems could be configured either in a "meshed" arrangement with all stations acting as up-links (transmitting to the satellite), or in a "star" configuration with stations linked by cable to a joint up-link station which handled broadcasting centrally.

- *Emergency back-up systems.* Back-up systems were security systems, which took over when cable networks broke down.

- *Very Small Aperture Terminals (VSAT).* This technology involved very small receiver stations, which could be mounted virtually on any building.

VSAT was dependent on the gallium arsenide technology introduced in down-links in the mid-1980s (this technology contributed to a 90% price reduction, while the requirements in respect of dish size at the same time decreased considerably).

1.1.4 The Beginning of Cable

One of the more surprising side effects of the development in the satellite market in the USA was that it stimulated the expansion of its own alternative: cable networks. In 1948 the American authorities stopped issuing new TV licences to prevent the limited airspace from being filled with national channels instead of local ones. As a result, many areas were not able to receive TV as soon as they had hoped and manufacturers were unable to sell their TV sets rapidly enough. This stimulated the expansion of cable networks, or "Community Antenna Television" (CATV). Transmission via cable had the benefit of not interfering with the limited frequency capacity in the air, and this was thus an ideal way to overcome the problem. These first cable networks had the sole purpose of offering TV to areas which, owing to their location, were unable to catch the national TV programmes by means of ordinary microwave antennas. Often the users could see the central receiver antenna on a nearby hilltop. A family typically ran these "mom-and-pop" operations with the patience to get up at regular intervals to find out why the network was not working in some location or other. The cable phenomenon gained somewhat of a reputation for being "some strange wires stuck together with chewing gum."

When the restrictions were lifted in 1952, the authorities believed that the further expansion of cable networks would come to an end, but this was

Table 1.1   Milestones in the opening up of the American commercial
                 TV market

1972	Open Skies policy adopted
1975	First pay-TV service delivered via satellite
1979	Satellite receive-only down-links deregulated
1980–81	Cable TV distant signal limitations lifted
1982	DBS satellites authorised

not to be the case at all. Cable networks had at that time already taken root
in the foothills of rural America, having reached critical mass, and they
were now ready for continued growth.

In the mid-1960s cable networks expanded beyond the rural areas and
were laid down in more and more urban areas. The operators of these
networks were typically more commercially oriented, better financed and
more professional. The first American cable operators running multiple
networks now appeared on the scene (Table 1.1).

When cable reached the suburbs it meant access not only to an additional
audience, but also to an audience with much higher average purchasing
power and who were often opinion makers, and to an audience that were
cheaper to serve because they lived closer together. This allowed for a
much greater commercial potential.

One way to meet this new potential was to redistribute satellite-based
channels. Initially the first cable networks had redistributed terrestrial
television only, but now they could add television channels that they
received via satellite. An increasing number of cable operators now
began to send professional door-to-door salesmen to private households
to entice people to take out subscriptions for cable connections. Suddenly
people could get access to as many as 30 TV channels!

This was also where the problem known as ''churning'' was first
observed. Cable customers were often referred to as ''the forgotten customers''. Once they had taken out a subscription they heard no more from
the sales department, and consequently many of them were soon cancel-
ling their subscriptions.

In 1986 there were 44 million cable subscribers in the USA. Typically a

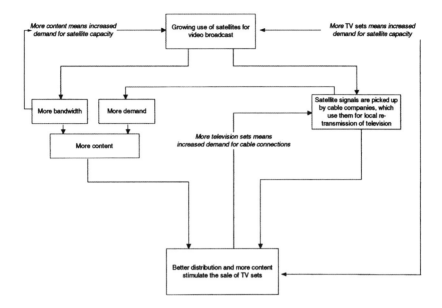

Figure 1.2   The satellite–cable growth cycle. The diagram above shows a typical example of inter-dependent technologies that stimulate each other

subscriber would pay US$ 2–5 a month for 12 TV channels. Customers willing to pay US$ 10–15 a month could get access to some 36–54 channels, and special pay-channels typically cost US$ 10 a month. The installation of cable networks was either free or would cost approximately US$ 25.

Interestingly, the tremendous success of cable was contrary to the plans and expectations of the authorities, which had not expected people to be willing to pay for a cable connection when they could get TV free with an ordinary antenna (Figure 1.2).

Although cable was originally an alternative to satellite delivery, the two technologies actually converged and became more and more interdependent:

- Satellites delivered channels to cable networks thus ensuring that many cable networks across borders had access to the same channels.

- The increasing cable distribution potential made it commercially feasible for media companies to create new channels.

- New channels enticed people to buy more television sets.

- New television sets increased the demand for satellite, cable and channels.

This was an early example of how technical convergence in electronic media and delivery could stimulate an increased diversity in the content offered.

1.1.4.1 Growth of Cable in Europe

While Europe had pioneered both radio and television, it was behind when it came to the introduction of cable TV. In the UK the so-called "Annea Report" had dismissed cable in a few paragraphs, calling it a "ravenous parasite" that should be confined to the roles of broadcasting relay and local community service. When the Conservative government took office in 1979, this view was altered, and in 1980 the Home Secretary authorised pay-TV experiments in various areas.

For many years the cable industry in continental Europe was also a rather sleepy affair. It was only in the 1980s that things began to change as cable penetration in Benelux, Germany and Scandinavia approached or even exceeded the American level.

1.1.5 The Emergence of Large TV Empires

The development of cable television had a major strategic implication: with the ability for continuous additions to the number of channels deliverable to each household it opened the market for more segmented media approaches. Some time after cable TV took off in the USA under the leadership of traditional television networks, for example CBS, NBC, and ABC, private TV empires started to develop their own niche market segments, such as news, children's programming, sports, weather, etc. The pioneers of each segment tended to obtain the dominant position in the respective market segment and thus became extremely difficult to challenge after a while, as even very small entrepreneurial start-ups like CNN, MTV and MBO were able to develop into large TV empires. This phenomenon was a typical example of so-called "first-mover advantage" in electronic broadcasting.

1.2 THE POWER OF BROADCASTING

The development of the broadcasting business has in many ways been amazing. Assume that anyone had told the inventors of television the following:

- There would be more than 1,300,000,000 units of their invention installed within 70 years.

- People would be able to watch up to 200 channels on it.

- It would become normal to have it turned on for 4–5 h a day.

- Some programmes on it would attract more than 1 billion simultaneous viewers.

- People in slum areas who could hardly afford proper food would save up for years to buy it.

- Politicians would have little or no chance of getting elected if they were not good at communicating through it.

They probably would not have believed it. But this is what actually happened, and what brought electronic broadcasting to its powerful position today was *the commercial power of sharing*. Broadcasting is based on a high-speed one-way connection that delivers the same programming into a very high number of households. This has four vital consequences:

- The high transmission speed makes it possible to deliver extremely compelling and emotional content with no waiting time whatsoever for the receiver.

- The concept of shared bandwidth makes it possible to deliver this content at such a low cost per user that much of the programming is in fact offered without any charge for the connection at all.

- The concept of large, simultaneous audiences makes it commercially feasible to produce programming that costs millions of dollars per hour – and yet makes money.

- The concept of large simultaneous audiences also creates a community feeling where people meet at work and ask ''did you see Jay Leno on television yesterday?''.

Broadcasting was not a bad idea.

New concepts using existing technology take off fastest

Close examination of growth rates for the major convergence phenomena reveal an interesting (but very logical) phenomenon. A new service/product offering that does not require major build-up of infrastructure will grow exponentially for a long period of time, while phenomena that require major investments are more likely to follow a linear growth rate. Examples of the former are the Internet, and examples of the latter are satellite and cable television.

2. The Internet

"While you are destroying your mind watching the worthless, brain-rotting drivel on TV, we on the Internet are exchanging, freely and openly, the most uninhibited, intimate and – yes – shocking details about our 'CONFIG.SYS' settings."

Dave Barry

Data broadcasting can perhaps best be described as the phenomenon where "the broadcasting business meets the Internet". So, while we have already looked at a bit of broadcasting history, we should perhaps now take a brief look at the background of its rapidly growing counterpart: the Internet. This evolution is just as amazing as the first development of broadcasting because of its tremendous growth rate and diversity.

2.1 THE CONCEPTION OF THE INTERNET

In 1964, Paul Baran of RAND Corporation (an American Cold War think-tank) published a proposal for a network structure that would be so rugged that it would continue to work even if America was under nuclear attack. All nodes in the network that he proposed would have identical status and they would be able to source, receive and send data to any other node through any route. If a direct route was inoperable, the system would have the ability to identify and find alternative routes. During the succeeding years the concept was discussed at MIT, UCLA and RAND. Finally the Pentagon decided to fund the development of a network called ARPA-NET, which was introduced in 1969 with four nodes. The network grew to 15 nodes in 1971, and 37 in 1972.

In Europe the Internet concept was implemented during 1981–83 for a project called STELLA, which aimed to connect local area networks. One of these was CERNNET, which connected networks in CERN (Switzerland/France) and Pisa (Italy), and the other was the Cambridge Ring that connected CERN and the Rutherford Laboratory.

In 1983 the military part of the American network was segregated from the rest, while more and more new networks connected to ARPANET, which was now using the public domain "TCP/IP" protocol. Within a year the National Science Foundation was involved in the development of the net, adding better networks and stronger servers. The Internet was now a well-established technology, but it was still neither developed as a medium, nor very attractive as a business.

2.2 GROWTH AND MATURITY

The year after the segregation of the military and the civilian part of the Internet, in 1984, Tim Berners-Lee at CERN wrote a proposal for what would later become the "World Wide Web". Berners-Lee's proposal contained a number of new ideas, like the concepts of URLs, HTTP, HTML, and browsers. When the first browser (Mosaic) was later developed at NCSA, the real commercial exploitation of the Internet began. It was now emerging as a medium and a business.

The next wave on the Internet came surprisingly fast. People soon found out how you could create graphical websites, use hyperlinks in smart ways, create virtual communities, etc. During the years 1994–96 it became clear that the Internet was becoming the fastest growing consumer offering, and business tool, in the history of mankind, and that it was rapidly maturing as a new and exciting medium.

The rapid growth of the Internet happened not only because of the sheer brilliance of the core technologies behind it; there were three other key factors.

The first of these driving factors was *core innovation* in information technology, such as the continued increase of computer power and an avalanche of new telecommunication technologies. Some of the innovation processes in core information technology lead to sustained *exponential performance growth*, a phenomenon that has been relatively rare in economic history. Two rules described these phenomena; *Moore's Law*, which stipulated that chip capacity would double every 18 months while its price fell to half. And *Gilder's Law*, which said that the total bandwidth of telecommunication would triple every 12 months. Any system that contains two such exponential processes is obviously prone to dramatic development. This dramatic development was best exemplified by an explosive growth of the number of *applications*, such as bots, search engines, avatars, agents, etc.

The second key factor behind the dramatic growth of the Internet was a new and accelerating trend towards *open standards*. Open standards as a concept had of course existed for centuries, but they did not take off in a major way in the IT industry until the early 1990s, as more and more managers understood their commercial potential. Open standards had many effects. One was that high-tech companies would move from vertically integrated models with proprietary end-to-end systems towards either more horizontally focused approaches, where each supplier would focus on serving smaller parts of the value chain, or towards a solutions approach, where companies would assemble solutions around open standards by working closely with other providers in the value chain. Another effect of open standards was *convergence*, which meant that applications were able to run on larger and larger numbers of different systems provided by different vendors.

The third independent driver of the growth of the Internet was a new, global trend towards *de-regulation* of telecommunications markets. This trend, which was driven by an increasing recognition of the advantages of free competition, was partly stimulated by international organisations such as GATT/WTO and the European Union. De-regulation meant that network infrastructure providers that previously had operated under a protected status suddenly met new competition from many angles. The new, open market space stimulated formation of many start-up telcos and a rush among new and old operators to launch innovative services and reduce cost and prices. The Internet turned out to be one of the main instruments in their new competitive battle, and an important effect was *falling prices*.

The combination of new applications (stimulated by core innovations), technical convergence (stimulated by open standards) and falling prices (stimulated by de-regulation), combined with the fact that the Internet used existing infrastructures, enabled the exponential growth in the number of Internet users and in revenues. This revenue stimulated again the frantic development of new core technologies and new applications – positive feed-back loops created a sustained growth process as visualised in Figure 2.1.

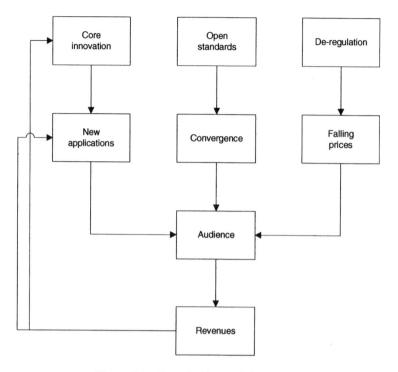

Figure 2.1   Growth drivers of the Internet

2.3 THE NETWORK EFFECT

Positive feed-back processes that create sustained growth is not unusual in economic systems. However, as it was soon discovered, there was something about the development of the Internet that was different from what has been seen in many other markets: a so-called "network effect". An example of a network effect: there would be no use in being the only person on earth with a fax machine; the more people that have the same device, the more valuable it became for each user. It was well known that network effects existed in situations where the value of having a communication technology increased proportionally with the number of other people that has the same technology. However, the interesting aspect with the Internet was that the value of a network that connected individuals with individuals increased not in *direct* proportion to the number of users, but in *exponential* proportion to the number of users. A network with 2 million users had a value to each user that was far higher than the combined value of two separate networks that had 1 million users each. This phenomenon was called *Metcalfe's Law*.

The number of people connected to the Internet grew at an exponential rate, and the value of being connected to each individual grew at an exponential rate. The result was that Internet data traffic doubled every 3–4 months. Nothing like this had been seen before. Not in technology, not in media and certainly not in any other business.

2.4 INCREASING RETURNS

Several economists have investigated economic phenomena that evolve in a digital economy. Two of those, Brian Arthur and Paul Krugman, were particularly interested in a phenomenon called "increasing returns". Conventional economic theory assumed that corporations had decreasing returns on investment (lower return on each new dollar that they invest). They would consequently keep investing until they reached a point where the marginal return on new investments become zero. However, what Arthur and Krugman noticed was that many providers of digital products, services and solutions experienced that the return on each new dollar invested in winning concepts was actually *larger* than the return on the previous dollar invested in the same concept. The main reasons were:

- The network effect (Metcalfe's Law).

- The fact that the marginal costs of replicating electronic deliverables could be minimal.

- The possibility for the largest players to become de facto standard providers.

So supply stimulated supply (increasing returns) and demand stimulated demand (network effects) – a truly explosive dynamic structure. And this was not even the whole story: while every economic boom creates inflationary pressure there were elements in the Internet boom that exerted downward pressure on inflation:

- The Internet created transparent markets, which intensified price competition.

- The Internet enhanced productivity by enabling disintermediation of middlemen, stimulating collaborative work approaches reducing time-wasting procedures and accelerating the exchange of software (and thereby its own continued development).

Low inflation meant low interest rates and high equity prices (which in some cases triggered momentum investment and bubbles, but this is an other story). This, combined with a demographic boom of middle-aged people saving for their pension meant access to an abundance of capital available for new Internet projects.

When big is beautiful

It can be argued that the value of a broadcast network grows in direct proportion to the number of users, while the value of the Internet grows in exponential proportion to the number of users because of the network effect. However, broadcast networks scale much better than Internet services in the sense that they have almost fixed costs and virtually no marginal costs related to the connection of new users. This is a reason why big also is also very beautiful in the broadcast business.

2.5 THE FUTURE OF THE INTERNET

Most products and services obey a few simple rules as they move from the introduction phase to growth and maturity. Textbook descriptions of some of these rules are shown in Table 2.1.

What these textbook models say is that products first address a few market segments, and later on many. They also suggest that products initially have simple and basic functions, but later on a rich diversity. And they point out that they initially are sold as stand-alone products, later as

Table 2.1   Traditional product development phases

	Introduction	**Growth**	**Maturity**
Number of market segments addressed	Very few	Some	Many
Quality and diversity of applications	Basic	Evolving	Rich
Convergence with other products and services	Stand-alone	Converged	Embedded

combined solutions that have converged with other products, and that they in the maturity stage often become an expected feature in other products, like the radios in our cars. These basic rules have applied to a very wide range of products in all sectors. Electronic products and services are no exception.

Take for instance computers. The first computers were stand-alone mainframes that were designed to address a limited number of market segments and to perform a limited number of tasks. This was followed by minis and PCs, which had much richer selections of applications. These new computers would typically converge with other devices. A PC could, for instance, function as a CD player, a type-writer, a calculator, a clock, a games console, etc. The maturity stage has brought computing further into a rich multitude of devices that we use every day such as toys, set top boxes, MP3 players, multimedia kiosks, and cameras, etc. The computer is in many cases no longer just a *converged* product: it is completely *embedded* into the other products that we buy. This embedding is even in many cases so discrete that the user is not fully aware of its existence. Few people know, for instance, that there are billions of very simple chips (so-call ''jelly-beans'') tracking heat, age, physical position, electrical flow, speed, pressure, and light in dynamic environments for us everywhere.

An interesting aspect of the traditional product cycle is that new products most often do not lead to the elimination of older concepts. The mini did not kill the mainframe. Nor did the PC kill the mini, the games console kill the PC or the jelly bean kill the games console. (The same has been the case for broadcasting, where radio did not kill books and television did not kill cinemas.) Each new wave simply ads to the diversity of an environment that increasingly resembles an ecosystem.

A similar process is happening for the Internet. It was initially a military concept and then a simple tool used by academics – mainly to exchange e-mails. The Internet is now used by a diversity of people for a multitude of purposes, and it is firmly embedded into the functionality of much PC software. It is not difficult to reach the conclusion that the Internet will continue to follow the typical evolution of any other group of products and services.

A basic rule for electronic media is that *convergence in core technologies leads to increased diversity in applications* (Figure 2.2). Let us take a quick look at what that actually will mean. We can start with market segments and application areas.

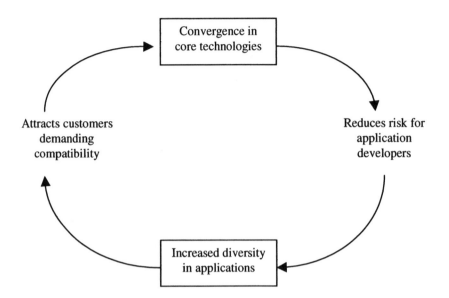

Figure 2.2   Innovation and diversity in computing, telecommunication and media

2.5.1 Ubiquity and Diversity

The large majority of the Internet traffic is still generated by people using a PC and a standard modem. While this approach is suitable for a multitude of applications and for a lot of people, there are even more people and applications that it does not suit. The future is likely to enable more people to benefit from more Internet applications through the use of new devices. These will include toys, watches, clothes, glasses, cars, phones, televisions, multimedia kiosks, elevators, vending machines, traffic lights, Personal Digital Assistants, microwave ovens, heating systems, pets, walkmans, wall-mounted mini cameras, LEGO robots, printers, games consoles and voice mail systems. Many of these devices will be operated by people, but more and more devices will simply communicate directly with each other without involving human interaction – and it may not be long before there are more things than people using the Internet.

Availability of an increased diversity of connected devices will mean demand for more and more applications and services. The person using a fast connection should, for instance, not get the same content as the one using a slow connection, and nor should a person using a phone or a car screen receive the same content as the one using a PC, etc.

There will also be a demand for different service levels. One problem that the Internet has today is lacking *Class of Service* (CoS). CoS means classification of different services so that they can get different treatment in the network. It could, for instance, mean that mission-critical business data is given priority over entertainment for teenagers. The Internet was not designed with that feature in mind. Another problem is lack of *Quality of Service* (QoS). Quality of Service has two dimensions: the ability to guarantee a certain bandwidth for a given application, and the ability to guarantee a minimum time between delivery of each package (something that is particularly important for audio and video applications such as Internet telephony and videoconferencing). Many businesses are choosing to build and maintain their own costly networks instead of using the Internet – exactly for these reasons.

2.5.2 Convergence and Embedding

One of the most exiting aspects of the development of the Internet is convergence and embedding. The simplest form of convergence is "bundling": you buy a computer and receive Internet software for free. A stronger version is when you buy a computer and Internet connectivity is an integral part of it. The strongest version would be that you buy a product that connects to the Internet all the time without your intervention. Internet connectivity has then been reduced from a product to a feature. Examples could be:

- Wristwatches that use a wireless Internet connection to synchronise with a scientifically managed world clock.

- Cars that inform the factory about car status and position (using Global Positioning System) via wireless Internet when the airbag inflates.

- Microwave ovens that receive electronic cooking instructions together with new recipes via the Internet.

- Elevators that alert the factory via the Internet when they need service.

- Light signals that call a technician when a bulb does not work anymore.

- E-toys that communicate with the television via wireless Internet and make funny comments to the people on the screen.

However, the single biggest area of convergence and embedding for the Internet will come from another angle: it will be the integration with the broadcast business. This is perhaps best understood if we examine yet another law. It is based on something called ''Amdahl's Constant''.

2.6 PROPORTIONS BETWEEN BANDWIDTH AND MEMORY

The law is simple. It says that most intelligent entities (computers and live organisms) have roughly equal amounts of memory (content) and data processing capacity or bandwidth. This means, for instance, that if there are 1 million bits of content stored, then the bandwidth is likely to be approx. 1 million bits/s: bandwidth can, in theory, transmit the equivalent to all the memory in 1 s. This one-to-one relationship is called Amdahl's Constant (Figure 2.3).

Think of media. Traditional radio and television provides lots of content, but with too limited possibilities to do anything with it. If on the other hand, there is plenty of memory but little bandwidth, then you can do a lot with the content – but there just isn't enough of it. A good example is a games console. You can manipulate the content in very impressive ways,

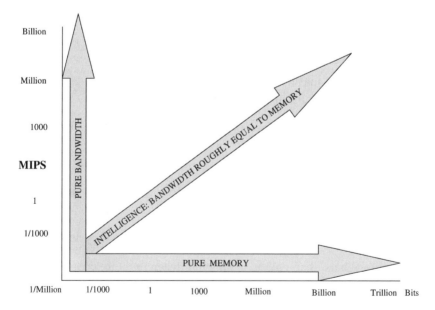

Figure 2.3 Bandwidth and memory. Intelligent systems tend to have roughly equal amounts of bandwidth and memory

but you only have the few games you have chosen to buy, and they do not change.

Hans Moravec, who is founder of the world's largest robotics program, at Carnegie Mellon University, illustrated this relationship in his book *Mind Children* from 1988, where he plotted the memory and bandwidth of viruses (the memory of a virus is the code in its DNA), bacteria, spiders, PCs, main frame computers, mice, monkeys, people and other animals and machines into an exponential graph that clearly illustrated the tendency.

Figure 2.4 shows examples of different machines and organisms and their combination of memory and bandwidth. It illustrates their rough compliance with Amdahl's Constant.

Now, anyone who has spent 45 min trying to download a plug-in from the Internet may ask if this Amdahl's Constant really applies there as well.

The Danish author Tor Nørretranders published in 1997 a book, *Stedet*

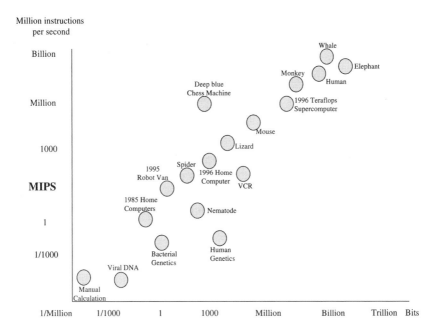

Figure 2.4 Amdahl's Constant. Source: Robot, Hans D. Moravec, 1999. The illustration shows numerous examples of intelligence. They tend to have roughly equal amounts of bandwidth and memory

som ikke er (The Place that Isn't), where he attempted to determine the relationship between all Internet content and all available Internet bandwidth. His conclusion was that the combined content on the Internet (websites, FTP files, newsgroups, etc.) was approx. 10,000 times as big as the combined effective bandwidth (measured as the highest combined backbone traffic). While the exact number is impossible to measure (and both numbers grow exponentially), it is clear that the Internet does not follow the norm: there is a bandwidth problem. Bandwidth is, in fact, the biggest problem.

Broadcasting networks, on the other hand, have massive bandwidth; not only for backbone transportation, but also to every single one of the billions of end-users. Anyone can verify the difference through a simple experiment:

• First you try downloading a music sample and video clip from the Internet

• Then you zap through the many channels that television and radio provides us around the clock.

The difference in speed is massive. A TV channel provides 30 full-screen pictures per second, plus stereo sound, and this appears instantly after the channel has been chosen. A cable TV network with 40 channels delivers $40 \times 30 = 1,200$ full screen images *per second.* Did you ever see that on a website? A digital cable network with 200 TV channels deliver 200 times stereo sound plus $200 \times 30 = 6,000$ full screen images per second. You get the picture.

A happy marriage between broadcasting and the Internet would have been extremely challenging when Internet was digital and broadcasting analogue, but we are now moving to a new scenario where both are digital. Data broadcasting is the technology that provides the bridge.

Table 2.2   Time-line of Internet events

Year	Event
1945	Vannevar Bush (Science Advisor to President Roosevelt) writes a proposal for a conceptual machine that can store information in such a way that users can create links between text and illustrations for future reference (the precursor of Hypertext)
1964	RAND Corporation makes proposal for rugged, nuclear war-resistant network system
1965	Ted Nelson coins the word ''Hypertext''
1967	Andy van Dam and others build the Hypertext Editing System
1969	ARPANET is launched with four nodes
1972	ARPANET reaches 36 nodes
1980	Tim Berners-Lee writes a notebook program that allows the user to make links between arbitrary nodes
1981–83	Internet concept is used at CERN to connect with other European networks
1983	The military part of the American Internet is segregated
1984	Tim Berners-Lee at CERN writes a proposal for what later becomes the World Wide Web. It introduces the concepts of URLs, HTTP, HTML, and browsers
1989	Tim Berners-Lee writes and circulates a proposal for information management containing descriptions of WWW structure
1990	First WWW software and browsers are developed at CERN
1991	CERN staff demonstrate WWW software at the 1991 Hypertext Conference
1991	CERN staff release first WWW browser
1993	By the beginning of the year there are about 50 HTTP servers world-wide. Several browsers are available
1993	NSCA release first version of Marc Andressen's ''Mosaic for X'' browser
1993	WWW traffic reaches 0.1% of NSF backbone traffic in March; grows to 1% in September
1993	In October there are about 200 HTTP servers world-wide
1993	The *New York Times*, *The Guardian* and *The Economist* cover the WWW and Mosaic

Table 2.2 (*continued*)

Year	Event
1994	Marc Andressen and some of his colleagues leave NCSA to form "Mosaic Communications Corp.", which later became Netscape
1994	The world's first WWW conference is held at CERN, Geneva in May. The conference has 800 applicants; only 400 are allowed in. A second conference, in Chicago, is held in October. This conference has 2,000 applicants, out of which 1,300 are allowed in
1995	CERN invites 250 reporters to learn to surf the Web during a 2-day seminar; 309 pupils from a local school teach the reporters how to use the Web.
1995	The Web Society is founded by the Technical University of Graz, CERN, the University of Minnesota and INRIA
1996	The number of Internet users is estimated at close to 30 million by the beginning of the year and is seen to grow by 15% a month
1996	The number of global websites passes 190,000 in May
1996	Various ISPs suffer extended service outages due to extensive traffic
1997	The domain name "business.com" is sold for US$150,000
1997	"Push" technologies and multicasting concepts are gaining ground
1998	Digital Corporation estimates the global number of web pages at 275 million. NEC estimates that it is 320 million
1998	Internet users are on-line judges as 12 world champion ice skaters perform live
1998	The concept of Internet "portals" gains ground
1998	Use of MP3 for music downloads take off
1998	Web TV, which enable people to access web pages from a TV screen, takes off in the USA following aggressive marketing campaigns
1998	A number of pilots of Internet2 services are launched
1998	NOKIA launched a telephone that enable users to connect to the Internet
1998	AOL buys Netscape for US$ 10.2 billion
1998	The market's talk of net computers fade as the focus shifts to a wider range of IP devices

Table 2.2 (*continued*)

Year	Event
1998	Data traffic overtakes voice traffic over telco networks in a number of countries
1998	British Telecom commences pilots of data broadcasting services over ADSL
1999	Yahoo buys Broadcast.com for US$ 5.7 billion
1999	"Portals" become the buzzword of the year as many of the original search engine companies evolve from pure search tools to media companies and then portals
1999	The Melissa virus invades computers world-wide
1999	CMGIs accumulated return since its Initial Public Offering in 1994 surpasses 27,000%. CMGI is a leading investor in Internet companies
1999	Barnes&Noble.com go public. Amazon.com reduces prices of best selling books 1 week before Barnes&Noble.com's IPO
1999	Shares of E-bay, the on-line auction house skyrocket and Amazon.com announce their own, competing service
1999	British Telecom commence pilots of data broadcasting services over satellite
1999	A range of new IP-enabled devices such as refrigerators, microwave ovens, PDAs, mobile phones, cars are launched
1999	MCI/Worldcom begins upgrading the US backbone to 2.5 gigabit/s
1999	Merill Lynch launches electronic trading services for its customers
1999	An increasing number of net start-ups start buying up old brick-and-mortar companies against shares
1999	The US government increase liquidity in anticipation of widespread problems concerning the millennium bug. The liquidity stimulates a strong surge in stock prices for technology, media and telecommunication companies (which is called the "TMT" sector)
2000	The millennium bug fails to materialise, except for a few places, where timekeepers list the first day of the millennium as "January 1, 1900"
2000	"Ubiquity", "broadband", "B2B" and "mobility" become the buzz-words of the year

Table 2.2 (*continued*)

Year	Event
2000	Several new B2B Internet exchanges (electronic market places) pop up every day
2000	Japan's NTT DoCoMo report huge success for their I-mode data enabled mobile phones
2000	AOL and Time Warner announce merger. This is seen as an indicator for the continued convergence between telecommunications and media companies
2000	Take-up of mobile phones continue to beat previous forecasts
2000	WAP phones are launched across Europe
2000	Amazon.com launches a series of new services as it prepares to become the ultimate electronic consumer mall
2000	The big music labels sue MP3.com for copyright violations – and win in a settlement deal
2000	New car communication systems combining digital broadcasting and web access are piloted in Germany and elsewhere
2000	Internet stock prices continue their rise until March, where they peak. A crash follows, and many companies fall more than 80% within a few months
2000	The "Lovebug" virus spreads across the globe within a few days. It causes billions of dollars of damage. Several copycat viruses appear during the following weeks, however, none are as damaging
2000	Microsoft is ordered to break up, and launches an appeal
2000	A number of high-profile Internet companies go bankrupt, and analysts predict that the majority of Internet companies will go bust within 1–2 years
2000	Bertelsman strikes a deal with Napster for music file-sharing with a revenue model for the content providers
2000	Worldzap and Sonera launches test service providing near-live video clips from Bundesliga football goals to PDAs
2001	The sell-off in Internet Stocks reaches its climax

3. Broadcasting Meets the Internet

"Everything that can be invented has been invented"

Charles H. Duell, Commissioner, US Office of Patents, 1899

We have already defined data broadcast as follows:

"Data broadcast is a concept whereby a combination of video, audio, software programs, streaming data, or other digital/multimedia content is transmitted continuously to intelligent devices such as PCs, digital set top boxes and hand-held devices where it can be manipulated. The broadcast concept means that although a return path might be available, it is not required. Content is received without being requested".

We shall in the following try to position data broadcast from two different perspectives: as a technology and as a business.

3.1 DATA BROADCAST: TECHNICAL POSITIONING

Telecommunication is a diverse business, but it can be divided into two very basic concepts:

- Point-to-point communication

- Point-to-multipoint communication

Point-to-point communication is converging around the Internet with its TCP/IP protocols. TCP/IP has a number of advantages, but it has one pain. It is slow – mainly for three reasons:

- *The acknowledgement process.* Every time you send an information package a message needs to go back through the network confirming that it has been received by the individual.

- *Strain on servers.* When 1 million people want the same information, the server has to send it out 1 million times and read all the package acknowledgements at the same time.

- *Strain on networks.* What is sent to John is not the same as what is being sent to Paul and Mary. So they cannot share the same infra-structure. Three users generally means three times the capacity requirements.

The pain of the Internet is, in other words, in the pipe.

The other main concept of telecommunication – point-to-multipoint – is what we looked at in the first chapter. This is broadcast, which today means mainly radio and television broadcast. Radio and television are *fast, cheap* and, you may say,*"for the masses"* – a satellite/cable-connected television receives massive amounts of content – 24 h a day. No dial-up required. No per-minute fee. Broadcast is clearly a strong commercial concept. But it is not smart. Radio and television are single-media – not multimedia.

So how do the broadcast world and the Internet world relate? Well, firstly it should be noted that while everyone talks about the Internet today, the reality is that most electronic content distribution is broadcast. If Bob has a cable TV subscription with 50 channels, then he may receive 750 Mbits of content per second – 24 h a day. Looking at one of those channels at a time for 4 h may provide him with 27,000 Mbytes of content. If Bill plays on the Internet for 4 h with an average content download speed of 20 kbits/ s, then he receives 36 Mbytes of content. Broadcast delivers more.

This is no coincidence. Broadcast is so much bigger because it is so much more efficient at delivering really popular content. There will, for instance in the future, be no point in trying to deliver widely requested video-based content as "video-on-demand" over the Internet if it can be broadcast instead. A converged medium would be able to handle that.

Data broadcasting can, from a technical point of view, be positioned as the technology that merges broadcasting with the Internet (Figure 3.1).

The phenomenal drive of the Internet and electronic commerce, and the introduction of digital television (and higher bandwidth mobile phones) are processes in the market that prepare the way for ultimate convergence, which will mean that:

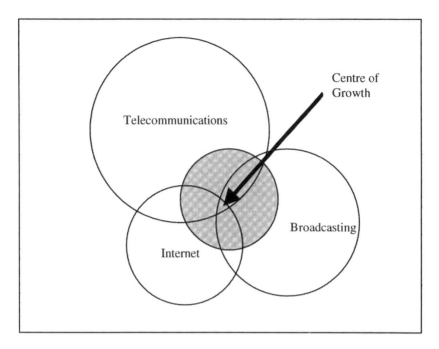

Figure 3.1 The convergence of telecommunication infrastructures, broadcasting and the Internet

- Internet content and data broadcast content from anywhere...

- is transmitted to a wide range of different devices...

- which we use at the office, at home, at public places and when we travel

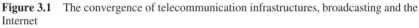

Music Choice Europe™: creating the three-layered medium

Music Choice Europe™, is one of the leading data broadcast pioneers. The first prototype of the Music Choice data broadcast service shows a typical combination of the three-layered data broadcast medium. The first layer in this data broadcast prototype is a number of non-stop digital music channels provided as MPEG streams over broadband. The user can listen to it (it is CD quality and involves no download delay) and can always read related information about the track, artist and album that is currently "on air".

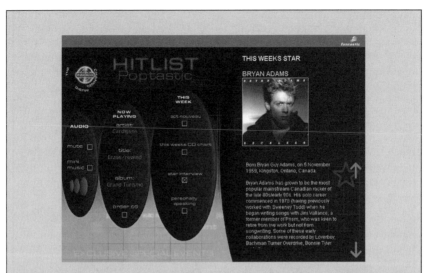

The second layer is the "walled garden". This is interactive content which has been downloaded to the hard disk (called "walled garden" since it is separate from the Internet). Music Choice Europe's™ prototype service provides new information to the walled garden whenever a new track starts. It works as follows: the "Intelligent Channel Compiler" on the server side receives information about a new track that is starting. It will then instantly and automatically link into a music e-commerce site on the Internet to seek information about purchase of the album. This information, and the order form, are immediately broadcast to all users.

The user will only see this order form if he likes the song and clicks to see the form. He can then add it to his electronic shopping basket and keep doing so over several weeks without ever dialling into the Internet (which is the third layer in the medium). However, whenever he wants to actually submit the order he has to connect into the Internet. The result is a very media-rich always-on service that provides ideal opportunities for the user to get to know new music and to buy it with maximum convenience. It combines the best of broadcasting with the best of the Internet.

- …and where we are able to mix the content from the external world freely with our own content.

Such a situation will of course be very attractive for the end-customer and for content providers, but it will also create what has sometimes been referred to as "strategic inflection". It was Intel's Chairman, Andrew S. Grove, who coined the expression "strategic inflection points" to describe situations where a business undergoes a dramatic change because a growth curve changes direction. Such a strategic inflection point is a situation where an external force has changed dramatically and where that change provides new threats or opportunities. Grove referred here to the "10 × factor": if a critical external variable changes by a factor of 10 or more, then there is a basis for strategic inflection.

Data broadcasting will typically provide on-line data into a computer or another IP device at more than 100 times (and up to several thousand times) the speed of a fast Internet connection, which means that it creates a strategic inflection point for the Internet business as well as for the broadcasting business. This strategic inflection point dramatically changes some of the commercial rules, since it provides new opportunities to operate companies "at the speed of thought" (Bill Gates) and to provide a powerful direct communication line between companies and consumers. The phenomenon will also provide strategic inflection for media companies and broadband network providers who are already directly involved in the distribution of bits and bytes. So from a commercial point of view, we can position data broadcast as the phenomenon that enables players in the value chains to provide fast and rich communication to their users anywhere, anytime and anyhow.

3.2 THE DATA BROADCASTING ECOSYSTEM

So data broadcast is a bridge technology between broadcasting and the Internet and is a new and extremely efficient means of providing fast and rich media. But for whom is data broadcasting relevant? Who are the players that will participate in this business?

Any business, including any new electronic media business, has a so-called "ecosystem" of suppliers. If we think of data broadcasting as an infrastructure to deliver packages and messages, then we can compare it to the ecosystem of traditional mail. The ecosystem of traditional physical mail includes:

- *Enablers:* people who provide enabling equipment like sorting machines, vans, stamps, etc.

- *Senders*: people who send information (individuals, advertisers, catalogue sales companies, etc.)

- *Distributors*: people who deliver the mail (bulk mail, direct mail, individual mail, registered mail)

- *Receivers:* end-users who receive the letters and packages.

The main ecosystem for data broadcasting is not dissimilar. It includes:

- *Enablers*: hardware enablers, software enablers, advertising metering providers, system integrators

- *Senders*: media companies, corporations/institutions, advertising sales agencies, shopping mall operators

- *Distributors*: media aggregators, physical distribution networks, service retailers, support providers

- *Receivers*: end-users who receive the digital packages.

The system may not appear particularly simple, though it is comparable with ecosystems not only for television but for any other media. Figure 3.2 illustrates the typical main tasks for each of the players in this data broadcasting ecosystem.

The data broadcasting ecosystem is similar to the ecosystem of other media businesses, and it is now in a process where different parties are trying to identify their roles.

Table 3.1 explains in more detail what each of the players in this value chain actually does.

It should here be noted that any player in the chain might choose to fill several of the functions within it. It should also be said that the data broadcast value chain is somewhat shorter with the launch of a corporate data broadcasting solution rather than for a launch to the mass market. Furthermore, a corporation, typically, will simply purchase the components needed from any of several competing vendors, which is relatively easy to do.

3.3 THE ORGANISATION PROCESS IN THE VALUE CHAIN

While an established ecosystem is taken for granted in a mature medium, it is in fact always a slow and painful process to organise as each new medium arises. Although each player in the system has significant

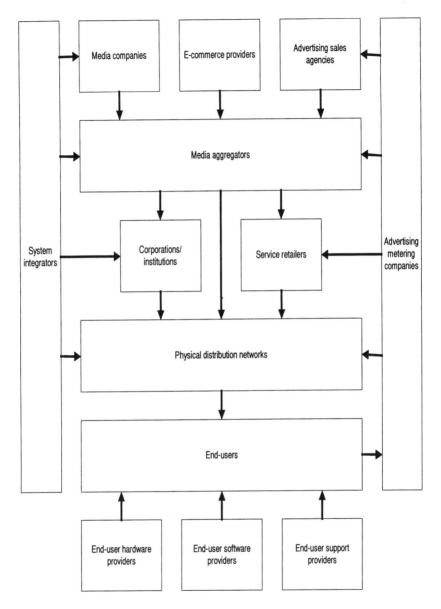

Figure 3.2 The data broadcasting ecosystem

interest in seeing the successful launch and take-off of a new medium, they will always undergo a process, which at times appears somewhat chaotic, in achieving a coherent co-operation procedure/flow.

Table 3.1   Typical main tasks for different value chain players

Main category	Sub-category	Main tasks
Enablers	Hardware enablers	Companies providing communication cards and encoders for DVB or other data communication. May also be PC or set top box vendors who may supply ''broadcast ready'' hardware. Can also include a hardware retailer
	Software enablers	Companies providing relevant software for aggregating, tracking, content broadcast scheduling, bandwidth booking, bandwidth management, and client content manipulation. Can also include retailers and value-added resellers
	Advertising metering providers	Companies who track and verify statistically the use of different components in branded media, including viewing of advertising, in order for advertisers to know the results of their efforts
	System integrators	Companies integrating data broadcasting solutions and channels into corporate, network and reseller environments in order to interface to, for instance, billing and conditional access systems
Senders	Media companies	Media brand companies providing data broadcasting content in order to earn directly derived revenues. Typically the content is provided either in incomplete format to content aggregators or in complete format direct to resellers

Table 3.1 (*continued*)

Main category	Sub-category	Main tasks
	Corporations/institutions	Corporations/institutions providing content typically in order to enhance their internal/external communication, improve data infrastructure, promote themselves, etc. The content is created by standard editing tools and then managed by data broadcast software tools. It is typically provided direct to end-users, and the corporation/institution is in charge of user administration
	Advertising sales agencies	Entities that sell advertising for the channels (in particular for branded channels)
	E-commerce providers	Companies operating electronic shopping malls which can be accessed from the data broadcast channels
Distributors	Media aggregators	Companies aggregating content from a number of different media companies in order to create a multi-channel package, widely through the use of software. They deliver the channels to the Network Operating centres of resellers and to corporations/institutions that wish to embed the content into their own channels
	Physical broadband networks	Networks providing bit stream services. Could be satellite, cable, xDSL (copper telco line), digital terrestrial or mobile. The networks are able to offer bandwidth booking with a flexible basis (time, bit rate, etc.)

Table 3.1 (*continued*)

Main category	Sub-category	Main tasks
	Service retailers	Entities who "own" the end-users for branded channels. These end-users may be Focused Affinity Networks, or they may be the public at large within a given territory. One may also view a corporation offering its own internal network as an end-user. The resellers are in charge of user administration, but report back to aggregators/content providers. The reporting may include statistical information, but not the exact identity of each end-user
	Support providers	Companies providing support for end-customers
Receivers	End-users	Local filtering and data manipulation

One of the reasons for this is that each group of players has a different technological framework as its starting point, therefore all the technological frameworks have to be interfaced initially for content to start flowing. The next chapter will provide an examination of the most important basic frameworks.

A factor in the increased competition among broadband networks

Broadband markets are rapidly moving from monopolies or near-monopolies to an era of all-out competition, as protective legislation is abandoned and new technologies allow competing broadband operators to invade each other's territories with competing services.

In time the decisive sales parameter will thus easily become price: if two suppliers offer essentially the same solution, then the market forces will tend to bring prices down to levels where not much money can be made by any provider. However, the broadband network provider can use adoption of data broadcast services as one of its ways of overcoming these problems, the reasons being:

• *Ability to segment the market.* Data broadcasting uses (unlike, for instance, analogue television) highly flexible bandwidth. Some content providers might transmit nothing for several hours, then go up in the Mbit range, then drop to a few hundred kbits, etc. Some might demand guaranteed real-time throughput (e.g. stock quotes), while others can tolerate limited buffering. Some might be happy to transmit overnight (e.g. database synchronisation), while others need prime time access. Some can plan their bandwidth usage well in advance while others need full flexibility. So being able to sell bandwidth with the flexibility that data broadcasting provides makes it possible to segment the market and create a win-win situation for customers and network operators alike.

• *Ability to provide value-added services.* "Value-added services" is always the strategic advantage necessary for any commodity business. How do you add to your offering so that price does not remain the only sales parameter? Offering data broadcasting can be a solution as it may offer anyone who connects access to broadcast content. Assume for a moment that there are news, weather, finance and music data broadcast channels available on one network while there is nothing available on the competitor's network. Which network will customers prefer to use? If the customer is a corporation that is interested in embedding any of these services into their own data broadcast channels, then they will prefer to use the network with the most compelling content.

• *Ability to attract new customers.* The third motive is simple. It is the ability to attract data broadcasting customers in the first place. The broadband network that does not offer value-added services will not attract new customers.The results of these three advantages of data broadcast for the broadband network are additional bandwidth sales, higher average bandwidth prices, reduced customer churn, and access to new revenue streams.

4. Technology Framework for Data Broadcasting Environments

"The wireless music box has no imaginable commercial value. Who would pay for a message sent to nobody in particular?"

David Sarnoff's associates in response to his urgings
for investment in the radio in the 1920s

The last part of Chapter 3 provided an overview of the data broadcast ecosystem. Many of the players in this marketplace are already involved in telecommunication and are using a range of well-established core technologies. These technologies provide the basic environment that data broadcasting technology and business models must interface with. This section will consequently provide a simple overview over these technologies so that the subsequent explanation of the specific data broadcast technologies and business models can be better understood.

4.1 MOVING PACKAGES – OLD AND NEW WAYS

Before telecommunication there was only one way to move packages and messages over long distances: physical mail. We all know what that means, so for the sake of simplicity we shall in the following sometimes compare some of the things that happen in telecommunication with physical distribution of letters and packages.

Let's start by representing the main steps that take place in an end-to-end communication example over an IP communications system and compare these to the old-fashioned physical distribution of mail.

As Table 4.1 illustrates, there are a number of "agents" involved in physical mail distribution – from the postman who picks up your mail, to the post employee who classifies it in the post office, and the lorry driver

Table 4.1 Comparing physical and electronic infrastructure tasks

Task	Solution for distribution of physical mail	Solution for electronic distribution
Preparing content for distribution	The content is put into envelopes or boxes	Dividing the content into electronic packets
Choosing delivery type and service	Classifying for Priority Mail, Registered Mail, Bulk Mail, and choosing the right delivery service	Choosing the type of service (HTTP, FTP, SMTP, RTP, etc.) and the transport mechanism (TCP, UDP)
Specifying sender and receiver and securing that any user is uniquely identified	Writing down receiver name and address. Maintaining a local area postcode system	Attaching addressing information. IP addresses assignment
Finding the right way to reach the destination	Automatic classification in the post office	Performing routing protocols such as RIP, OSPF
Choosing a vehicle for transportation	Using cars, boats, planes trucks, bicycles, conveyor belts and walking mailman to move the packages	Using different link types like Ethernet, ISDN, Frame Relay, ATM, MPEG-2 TS, etc.
Choosing a physical path	Selecting a route that the mailman will follow	Transmitting through fibre optic, coaxial cable twisted pair, satellite, microwaves, etc.
Providing security	Sealing envelopes, requesting signature and proof of identity upon reception	Using digital rights management tools, including conditional access, encryption and authentication

who delivers it to the destination. The same is the case in electronic content distribution – or "data communication". Different agents, here called *protocols,* are performing various activities to make the whole process happen. The combination of all protocols involved in the process is alluded to as a *protocol stack*, as each protocol will deal with the data at a different level.

Let's see in the next sub-sections how this protocol stack works in an IP data communications scenario.

4.2 DIVIDING CONTENT INTO PACKAGES

In digital data communications, when an information item is to be transmitted, it is usually divided and structured in small pieces called *packets* (equivalent to physical envelopes or packages). Each packet has information about the destination, the type of transport required and the information item it belongs to, so that are can be handled independently, and so that they can perhaps even follow alternative routes to the destination. The receiving end will sort all packets, check integrity and conform to the information item exactly as if it was at the transmitting end.

This approach is known as *packet switching* communications, versus the traditional *circuit switching* communications used for instance for analogue telephony. Such packet-based communication offers a more flexible and efficient way of communication, for instance:

- *It allows use of alternative paths.* In circuit switching all the information has to be handled by the same agent and follow the same reserved path. Circuit switching is in this connection equivalent to a situation where only the postman who gave you the letter knows where it is bound. He has to take it all the way to the final user. Packet switching, on the other hand, equates to putting an address on the letter so that it can be passed from agent to agent through the chain and still reach the final receiver.

- *It facilitates error correction.* Data communications provide different mechanisms to detect (and in some cases even correct) errors produced during the transportation of the information. Imagine that you are transmitting the complete works of Shakespeare, and an error is detected when it reaches its destination. The whole thing should be retransmitted again. However, if the data is structured into packets, then only the corrupted one should be retransmitted. This facilitates situations where many simultaneous communication users can share the same transport resources. As each packet has its own ID information, they will be differentiated at reception.

However, packet switching has also disadvantages. One is that each protocol has to provide a *header* with information about the protocol itself. The inclusion of these headers means additional data to transmit and this again means more time and resources required. There is a compromise between the size of the packets and the efficiency of the system. If packets are too small, we may be transmitting more header information than payloads. If they are too long, then we would have to re-transmit too much each time a single bit was lost (if the packet size is equivalent to the complete works of Shakespeare, then we would have to re-transmit the whole thing if just a single bit was lost in the first attempt!).

However, IP communications is packet based, so we will be alluding to this term very often during the text. We will focus on data broadcasting using digital packets rather than analogue streams.

4.3 CHOOSING A DELIVERY TYPE

Imagine that you are going to the post office with some letters. When you get there they ask you if you want to use "Regular Mail", "First Class Mail", "Registered Mail", "Bulk Mail", etc. We can call this the "delivery type" or "service". IP communications also provide multiple types of information *delivery types* and *services*.

4.3.1 Delivery Types

The main categories of delivery types for IP based communication are as follows:

- "FTP" (File Transfer Protocol) is used for file transference between computers (and other IP devices) on the Internet. By writing "FTP Remote Machine Name" we log into the FTP server and are able to transmit or retrieve files to and from the remote computer file system.

- If we want to seamlessly navigate through a particular information topic on the Internet, "HTTP" (HyperText Transfer Protocol) also retrieves files from web servers on the Internet, but in a very user-friendly way. HTML files are formatted and displayed with any contained multimedia elements by the web browser upon reception without caring exactly about how many files are retrieved, where they are stored or if they come from different machines.

- ''Telnet'' is the protocol used for *terminal emulation* over the Internet. The Telnet program allows the user to enter commands that will be executed in the remote machine as if they were entered directly on the server console. To start a Telnet session, a login on the server machine with the appropriate *username* and *password* has to be entered.

- To deliver messages that do not have to be available to the user immediately we may use the popular e-mail. The user will just leave the message in the ''SMTP'' (Simple Mail Transfer Protocol) that can be sent to the destination SMTP Server afterwards, and can be picked up by the addressee anytime afterwards – and in principle from any location.

- If we want a real-time communication with another user in the Internet we use ''IRCP'' (Internet Relay Chat Protocol) and the text we write will often appear immediately on the remote screen.

- We can go even further and send not only real-time text but also multimedia elements like audio and video with RTP (Real-Time Transport Protocol), that transmits real-time data, like audio and video, through the Internet. However, RTP itself does not guarantee real-time delivery of data; it just provides mechanisms for sending and receiving applications to support streaming data.

4.3.2 Delivery Services

Let's again go briefly back to the world of traditional mail. Once we have decided what delivery type we will use in a given situation, we have to make sure that the transporter or Delivery Service that we subscribe to supports it. For instance, we might not get a pick-up and overnight delivery from the national Postal Service, so perhaps we need to use a private Delivery Service for this service. Or perhaps we want to do a bulk mail distribution. A similar situation exists within IP communications, where we have two basic mechanisms for transportation, TCP and UDP, which support different categories of services. So what do they do?

The traditional Internet uses TCP, so when people refer to ''TCP/IP'', they are talking about TCP to an IP device (like a PC). TCP uses ''positive acknowledgement with re-transmission'' mechanisms. This means that whenever packets are sent to the receiver there are acknowledge messages going back to the sender. The sending computer reads these acknowledge messages (called ACKs) all the time and re-transmits whatever was not received correctly by the receiver. This procedure ensures that no data is lost, duplicated or out of sequence. A connection oriented transport proto-col like TCP offers obvious advantages, but also includes some disadvan-tages. The biggest disadvantage is that the same ACK mechanism that guarantees delivery and controls throughput may at the same time slow down the transport speed itself, firstly because the sender will be unable to transmit any more data until he gets some recognition on the appropriate reception of the portion that has already been sent, and secondly because it sends everything individually to each receiver. This makes complete sense if each user wants something different, but it is very impractical when many want the same. Also, this approach will always require a bi-direc-tional network and is clearly oriented to a one-to-one communication. This is not always a problem, but there are many situations where it is. Three examples:

- Telephone lines are still expensive in many countries.

- Bi-directional networks (like telephone lines) are typically providing much lower bandwidth than networks originally designed to be unidirectional (like satellite and cable television networks).

- The receiving device needs to transmit all the time that it receives. This may not be a problem for a PC, but it is for a mobile phone. Mobile phones use a lot more battery power when they transmit than when they only receive (the battery of a typical mobile phone may, for instance, support receive-only for 200 h but transmit for only 4 h).

The alternative for transportation on the Internet is the User Datagram Protocol (UDP), a best-effort connectionless protocol (transmit-only). UDP just keeps on sending out data without expecting any feedback from the receiver(s) on the status of the reception. UDP will never know if the data successfully reached the destination or not. On the other hand, as it does not have to wait for any response from the receiver it is very well suited for point-to-multipoint distribution (true broadcast) as well as for real-time streams. UDP is the typical transport choice for IP broadcast services – and for data broadcast.

The choice between TCP and UDP has implications for the services that can be provided. While TCP is able to "guarantee" end-to-end delivery and throughput control (but not how fast the data gets there), UDP provides a "best-effort" (a nice way to say that there is no guarantee of success) transport mechanism that guarantees speed, but not error-free delivery. So communication types like HTTP and Telnet will only be delivered by TCP while others like RTP can be delivered by UDP. However, as we will see when we get to the specific data broadcasting technologies, there is a bridging solution. It is, for instance, possible and often most feasible to use UDP to deliver packages to many and then TCP to deliver a smaller number of files to the few who did not receive all contents of the packages correctly during the first transmission.

4.3.3 Quality of Service (QoS)

We have already mentioned the lack of Quality of Service on the Internet in Chapter 2. The Internet's TCP can shape the throughput of the transmitter and guarantee that the data reaches the other end, but it cannot guarantee a predefined throughput value, as this will depend on the feedback coming from the receiver end. UDP can guarantee a throughput, as it does not wait for any acknowledge message to keep on transmitting. However, it cannot guarantee that all the data reaches the destination.

Some services like multimedia data distribution may require a predefined QoS in terms of:

- *Guaranteed throughput* or amount of data processed (transmitted/ received) per second

- *Controlled delay* or average time for a packet to travel from origin to destination (also called "latency")

- *Controlled Jitter* or difference in delay from one packet to another (this is important for real-time streams)

Nowadays this is not possible through the Internet. However, this can be achieved by alternative ways of IP data transportation, like data broadcast.

4.4 SPECIFYING DESTINATION AND GUARANTEEING A UNIQUE ADDRESS

Before sending out any piece of information we have to specify who is intended to receive it. In postal mail, we might use postal addresses like:

"Villa Fairhill, Ocean Boulevard 15, Atlantis, Utopia"

On IP communications we use *IP addresses* like:

"195.65.5.254"

An IP address is a 32 bit number (or four numbers between 0 and 255 separated by dots) that uniquely identifies a host on an IP network. On a closed IP environment like a private network, any IP address can be used, as long as it is unique in the network (where the first byte defines the type of addressing). However, for computers connected on the Internet there are various organisations that assign IP addresses on demand to guarantee their uniqueness. With the fast growth of the Internet, this range of addresses is starting to be too small and new addressing schemes to extend the traditional IP addresses will be launched.

There is a nice service for IP networks called DNS (Domain Name Service) that provides a translation of alphabetic names, which are easier to remember, into IP addresses. For instance

"intel.com"

will be translated by DNS into

"195.65.5.254"

When someone asks for an IP address (or a range of them) they can also ask for the reservation of a *domain name* that will be associated to it on the DNS.

4.5 PROVIDING INFORMATION ABOUT HOW TO REACH THE DESTINATION

The postman collects letters from the mailbox and takes them to the local post office, where they will be classified in the right output tray: for local distribution, inter city, or international, depending on the destination address written on the mail. In the last two cases, they will be delivered to another post office, closer to the destination, where they will follow the same classification process. This process will be repeated until each letter gets to the post office closest to the destination and can be delivered to the intended receiver.

This multi-hop mechanism to find the right route from origin to destination is similar to what is used on the Internet to route IP packets from origin to destination. The intermediate agents in charge of the classification are called *IP routers*, as they will decide the next hop depending on the IP destination address they find in the packet. Protocols like RIP (Routing Internet Protocol) or OSPF (Open Shortest Path First) are used in the Internet to exchange routing information among routers.

4.5.1 IP Multicast

Any piece of information of any type to be sent out by any media can have three potential types of audience:

- Exclusively one receiver: unicast

- A selected group of receivers: multicast

- All the potential receivers: broadcast

Unicast is what you would call point-to-point communication. It would in physical mail be equivalent to sending a letter to your aunt in Chicago. *Multicast* is point-to-multipoint communication. It would be like direct mail ("send the same to this specific list of addresses: xxxx"). *Broadcast* is also point-to-multipoint communication. It would be like when you send the same brochure to everyone in a given community ("send to all in area code xxx"). Multicast is, in other words, a restricted form of broadcast and is thus a part of what we describe in this book, Unicast on the other hands, is not broadcast.

There are similar choices in IP communications. Information can be sent on a point-to-point basis but it is also possible to adopt a point-to-multipoint approach. (IP addressing reserves a range of addresses to IP multicasting 224.0.0.0–239.255.255.255.) Point-to-multipoint IP multicast addresses do not belong to a particular terminal, but to a group of them, called *IP Multicast Group*. IP multicast groups are dynamic; meaning that any user can join and leave the multicast group at any time.

With IP multicast, a single packet can have multiple destinations and is not split up until the last possible moment. This means that it can pass through several routers before it needs to be divided to reach its final destinations. Or it can bypass all routers, which means that it is never split at all. Both scenarios lead to much more efficient transmission and also ensure that packets reach multiple destinations at roughly the same time. IGMP (Internet Group Management Protocol) is a protocol used to dynamically build routing information on how to distribute multicast data. IGMP messages are transmitted in the opposite direction to the wished multicast path.

The problem with unicast

Imagine that 100 people want to receive the same broadcast content via the Internet (unicast). This might, if it is a limited, highly compressed feed, require a total bandwidth of 2 Mbps (approximately 32 times a traditional modem speed). Providing a 2 Mbps feed into the Internet is expensive, but in no way prohibitive. Let's now instead imagine that there are 1,000 users. The bandwidth requirement has now risen to 20 Mbps, which is twice as much as the typical 10 Mbps Ethernet connection used in most local area networks. Now, assume instead that the content is really popular: 100,000 people are interested in this content. The bandwidth requirement is now increased to 2,000,000 Mbps (2,000 Gbps). One hundred thousand simultaneous sounds like a lot but not if you ask a TV producer. Some TV programs reach millions or in fact billions of users at the same time. One billion times 2,000 bits/s are 2,000,000,000, 000 bits/s. Problem!!

The only viable solution for simultaneous broadband delivery to very large audiences becomes multicast or broadcast.

4.5.1.1 New Aspects of IP Multicast

IP multicast can be used over any one-way infrastructure (such as a television satellite), but there are also developments underway that make it feasible to use it over the Internet's two-way infrastructures. While the Internet as a whole is not currently multicast enabled (i.e. not all routers support IP multicast routing protocols) multicast-enabled subnet islands have been connected together through ''tunnels'' to form the *Mbone*(Multicast Backbone). Tunnels simply encapsulate IP multicast packets inside TCP and send point-to-point to the end of the tunnel where the multicast packet is de-encapsulated. Mbone enables distribution and access to real-time interactive multimedia (data, graphics, audio, video, etc.).

Another interesting development is the *Reliable Multicast Protocols*. Examples of this are LGMP (Local Group Multicast Protocol) and RMTP (Reliable Multicast Transport Protocol). The latter has already been tested on the Mbone.

Table 4.2   Advantages and disadvantages of traditional point-to-multipoint alternatives on the Internet

	Traditional Internet	Internet Push	Multicasting/ broadcasting	Reliable multicasting/ broadcasting
Distribution	Point-to-point	Multiple point-to-point	Point-to-multipoint	Point-to-multipoint
Data delivery initiation	Receiver	Receiver	Sender	Sender
Communication protocol	TCP/IP	TCP/IP	UDP/IP	UDP/IP, TCP/IP
Requires return-path	Yes	Yes	No	Yes
Key advantage	Guaranteed data delivery	User transparent	Bandwidth efficient	Bandwidth efficient, guaranteed delivery
Disadvantage	Bandwidth inefficient	Bandwidth inefficient	Not widely available. No guaranteed delivery	Not widely available

It should be outlined here that not all situations of point-to-multipoint IP data distribution on the Internet mean that IP multicast is involved. For instance services known as *Internet Push* are based on servers that send information to multiple subscribed clients periodically. Also e-mail distribution lists send a copy of e-mails addressed to each member of the list (you might recall the ''lovebug'' virus!). In both cases the information was transmitted, as many times as there were receivers. This is, from a media perspective, a point-to-multipoint solution, but not from a technical perspective. It is consequently not the same as data broadcast, and it creates technical bottlenecks that data broadcast avoids.

Table 4.2 compares the alternatives for point-to-multipoint distribution that we can find on the Internet.

4.6 GETTING A VEHICLE FOR TRANSPORTATION

We have now looked at packages, addresses and destination information for the Internet and have in each case compared what we saw with a well-known concept from old-fashioned physical mail distribution. So let's move on to the next basic concept, which is to find a vehicle for the distribution.

Let's start with a slightly odd example: you have some documents to deliver to the other side of the Atlantic, and you only have a car... you are going to have a hard time! You had better give the package to an International Delivery Service that will pick it up with a van, put it on a train to the airport, deliver to the destination local post office in a truck and deliver to the receiver's door on the postman's motorcycle.

In data communications, IP enables the equivalent of this International Delivery Service, as IP packets can be transported over many kinds of different networks that ultimately link origin with destination. These linking networks are the transportation vehicles for IP.

One issue to consider here is QoS. We might be able to predict a flight duration or train journey fairly well, but the delay of a van in a traffic jam is quite unpredictable. In exactly the same way, some of these linking networks that will transport IP packets will have a predictable behaviour (ATM) while others will only be able to offer a ''best effort'' service (Ethernet), and this is one of the main reasons why it is so difficult to achieve QoS on the Internet.

Another interesting issue to consider is that this linking of networks will raise separate addressing issues. For instance LANs like Ethernet use the MAC address (known as hardware address, as each NIC (Network Interface Card) or network interface is intended to have a unique one). The address of the linking network changes on the packet as it travels through different links; however, the IP address will remain the same all the way (since it specifies the final destination). IP communications also provide some mechanisms to map IP addresses on link level addresses. An example is ARP (Address Resolution Protocol) which is used to link MAC addresses with IP addresses on Ethernet environments, in a dynamic and transparent way.

So we talk about linking between a number of different networks – that our little electronic package may rush from one network to another and then another again at the speed of light as it travels from sender to receiver. But which are these networks? Here are some examples:

- LANs (Local Area Networks) will transport IP packets through, local, usually small, environments on its way to the long distance network. Well-known examples are Ethernet, Token Ring, FDDI, etc.

- WANs (Wide Area Networks) will transport IP packets over long distances. ISDN, Frame Relay and ATM are common implementations. It should here be noted that ISDN represents a different class of communication than Frame Relay and ATM. ISDN is a switched circuit established to transport packets, whereas ATM and Frame Relay are packet networks themselves (to be accurate, frame and cell based networks, another "flavour" of packet networks) that can be used to transport data, voice, video, and any kind of information. This information can include data already packetised, like IP data. ISDN allows bi-directional connections of $N \times 64$ (or 56) kbps. ATM and Frame Relay allow configuration of shared or dedicated capacity of any speed. ATM establishes virtual circuits between source and destination according to predefined characteristics that will be guaranteed during all the duration time of the connection. In other words, ATM does support QoS levels, being able to guarantee bandwidth, latency and jitter and so making it suitable for any type of data transport service. However, ATM is a very point-to-point oriented network. Although it can support certain multicast applications, the result may not be very flexible or efficient; for instance, multicast groups are arranged by the information sender, so it might not be so easy to subscribe/unsubscribe to them.

4.6.1 When IP Gets a Trip Companion

Another not so explored possibility of transporting IP data is together with broadcast information, typically TV. There are two main ways to do it.

- In *analogue* TV, the VBI (Vertical Blanking Interval) is the part of a television transmission signal that is left clear of viewable content, to allow time for the television's electron gun to move from the bottom to the top of the screen as it scans images. This "free space" can (incredibly as it may sound) be filled in with some modulated digital information that will be transmitted seamlessly with the TV signal without affecting it at all and will be interpreted and used in the receivers capable of using it. The VBI has for many years been used to transport widely popular Teletext data throughout Europe (and closed caption text in the US) and is already starting to be used also for IP data transportation. Although the throughput per channel is not very high (it could, for instance, be 128 kbps for one television channel), the aggregate of all channels will mean an important constant flow, being at the same time a simple way to share already existing broadcasting infrastructure.

- In d*igital* TV, MPEG-2 Transport Stream can be used to transport IP data, but this time the whole capacity of a TV channel (4–6 Mbps) can be used for data broadcasting. (It is interesting to note this bandwidth for a TV channel when you are not trying to compress the content for Internet delivery. A single TV channel requires orders of magnitude more bandwidth than any normal Internet connection would support.) The Moving Picture Experts Group (MPEG) refers to a family of standards used for coding audio-visual information in a digital compressed format. MPEG-2 is mainly aimed at digital video broadcasting. It offers high resolution and CD quality audio and is used for HDTV (Higher Definition TV). MPEG-2 specifies not only the codification algorithm for the video signal, but also the transportation vehicle for that over any broadcast network. This transportation vehicle can be used to transport IP data seamlessly together with video signals through TV broadcasting platforms.

Alternative MPEG standards

We have already mentioned MPEG-2, but there are other important MPEG standards.

One is MPEG-1, which is used for storage and retrieval of moving pictures and audio on storage media with resolutions of 352 × 240 at 30 frames/s, slightly lower than standard VCR. MPEG-1 was optimised for CD-ROM or applications at about 1.5 Mbps.

Another is MPEG-4, which will be the standard for multimedia applications, gluing digital television, interactive graphics applications (synthetic content) and the World Wide Web (distribution of and access to content) and will provide the standardised technological elements enabling the integration of the production, distribution and content access paradigms of the three. MPEG-4 can scale up and down to adopt different bandwidth environments and this makes it very suitable for handheld devices.

The third one that should be mentioned is MPEG-7, which will specify how to represent the information about the information contained in a multimedia file to allow, for instance, efficient search for multimedia content using standardised descriptions in digital libraries, multimedia directory services, broadcast media selection, etc.

4.7 CHOOSING A PHYSICAL PATH

Trains will (hopefully) always keep on their rails and planes in their air routes. However, cars, trucks and motorcycles can go over motorways, roads, byways and city streets.

A transportation vehicle can be more or less flexible, but it always needs a physical road to travel. The main roads and railways in data communications are fibre optics, twisted pairs, microwave links, satellites and coaxial cables. The characteristics of each physical path will also be differentiated, some being able to transmit at high speeds with scarce errors in very long distances, and others capable of limited speeds and distances to keep errors under control.

One issue governing the physical path is that the infrastructure that is ideal for long distance transportation may be less so for the ''last mile''. You might say that it would be nice to have the plane from Swiss Cargo landing

Table 4.3 Overview of physical distribution networks for data broadcast

Main category	Sub-categories	Examples of core topologies and standards
Wireline	Enhanced copper	xDSL
	Fibre	Fibre-in-the-loop (FITL)
		Fibre-to-the-home (FTTH)
		Fibre-to-the-business (FTTB)
		Fibre-to-the-neighbourhood (FTTN)
		Fibre-to-the-curb (FTTC)
Cable	Fibre/coax	Star/Tree Hybrid Fibre/Coax (FITL, FTTH, FTTB, FTTN, FTTC)
Digital terrestrial		MMDS/LMDS
		MVDS
		WLL
		DVB
		DAB
Mobile		HSCSD
		EDGE
		GPRS
		UMTS
		DAB
		DVB
		WAP
		I-mode
Satellite	Geostationary Satellites (GEOs)	VSAT
	Medium Orbit Satellites (MEOs)	DBS
	Low Orbit Satellites (LEOs)	DTH
		DSR/ADR
Home networks	Wirebased	Bluetooth
	Wireless	IEEE 802.11
		HomeRF
		IrDA

in your garden for fast and efficient delivery of your mail, but then again –
perhaps not so nice... The same occurs with data paths. Some brilliant
technologies for long distance transportation are not so suitable for the *last
mile* to the receiver user. In fact, it is this *last mile* provided by *access
network* that has been the driving factor for data services deployment in
recent years. As the access network is the last step on the way of the data,
it will determine the quality of the received transportation service, whatever technology was used in the *backbone* or long distance network. Table
4.3 shows the main technologies considered for the last mile at the
moment.

Let's go through the most important of these network categories to see
what they really are about.

4.7.1 Wireline/Copper Pair

Copper pair is the most widespread access infrastructure in the world
(more than 700 million copper lines installed world-wide). It is also
known as *twisted pair* since it consists of two separate bundles of copper
wires, each of which is twisted and encapsulated in plastic.

Telco companies, with the purpose of providing telephony, an analogue
and narrow-band service, installed this infrastructure in many cases many
years ago. This infrastructure has for years been used for digital communications as well (such as connecting to the Internet) by using the well-
known *analogue modem.*The typical speed of such devices is limited to
about 56 kbps. The current modem speed is not the end of the road,
however. Depending on its status, quality and length, copper lines in
most cases can be capable of transmitting high-speed data, by installing
in both ends of the local loop special devices known as xDSL (Digital
Subscriber Loop) modems.

The term xDSL refers collectively to all types of Digital Subscriber Lines,
the two main categories being ADSL (Asymmetrical DSL) and SDSL
(Symmetrical DSL). Other types are HDSL (High Bit DSL), RADSL
(Rate Adaptive DSL) and VDSL (Very High Bit Rate DSL). Bellcore
originally introduced DSL in 1989 as a technology that could be used
primarily to provide video-on-demand over traditional copper-line.

ADSL technology has the ability to support *simultaneously* voice,
content-rich data, and video applications over the installed base of

twisted-pair copper wires with speeds as much as 32 Mbps for down-stream traffic, and from 32 kbps to over 1 Mbps for upstream traffic, so it is perhaps no wonder that people in the industry sometimes talk about "turning copper into gold".

ADSL is the most widespread xDSL solution. It typically offers 1.5–9 Mbps from source to receiver and 16–800 kbps from user to source. HDSL delivers 1.544 Mbps (T1 service) in both directions.

4.7.2 Wireline/Fibre Optics

Fibre optics lines are made of glass or plastic and are smaller and cheaper than copper. They use light pulses rather than electronic pulses for communication, and they have substantially higher bandwidth potential than copper over much longer distances. Fibre is mainly used in backbone networks but will gradually get into homes. The fibre optic can get more or less close to the end user. This is some of the terminology used to express alternatives in this topology:

- Fibre-in-the-loop (FITL)

- Fibre-to-the-home (FTTH)

- Fibre-to-the-business (FTTB)

- Fibre-to-the-neighbourhood (FTTN)

- Fibre-to-the-curb (FTTC)

Typical transmission rates over fibre range from 51.84 Mbps to 2.488 Gbps. Needless to say these speeds can pretty much move the bottlenecks from networks to servers. However, the actual deployment of fibre is still mainly in backbone networks – not anywhere near the final user.

4.7.3 Cable Television

Subway railways are fast, capable and efficient, but do not offer much flexibility for new routes, and are limited to local environments. This can be the equivalent for "CATV" networks in digital communications. Cable television is the second most widespread access path to the end user, although satellite is closing in. Cable networks were designed to distribute TV in local environments. As TV has very rich broadband content, these networks were designed to broadcast huge quantities of information.

The cable wire that goes into each home (or office) is almost always coaxial. Coaxial has one wire in the centre, which is encapsulated in plastic and surrounded by another conductor. This structure allows much higher bandwidth over longer distances than copper pair (effective throughput in coaxial can be 20,000 times that used in standard copper). However, it should be made clear that cable capacity is shared among all users in each branch – cable was not initially built for switched networking also, though it is possible to add switching to cable networks (this is in fact being implemented, or has been implemented, in many cable networks now).

The typical cable network topology (the backbone infrastructure) consists of a star of fibre optics from the "Head End" where TV signals are created and aggregated, and a tree/branch coaxial cable from the terminating point of each fibre. Depending on the design of the network, the fibre may terminate more or less close to the end user, producing again the scenario of FTTL, FTTC, etc. where the last mile is coaxial cable instead of copper pair.

Traditional cable networks are designed with one-way transmission in mind. However, cable television operators world-wide are now upgrading their cable network, so that they are able to support two-way communication as well.

A typical hybrid fibre–coaxial solution with cable modems will have an upstream bandwidth of 500 kbps to 4 Mbps capacity and downstream transmission speeds of 30 Mbps.

4.7.4 Digital Terrestrial

Terrestrial wireless is a communication infrastructure based on point-to-multipoint microwave distribution. The receiving device for terrestrial wireless may be either mobile or stationary.

"MMDS/MVDS" (Digital Multichannel Multipoint Distribution/Video Services) are the most popular implementations. These are also alluded to as "wireless cable", as they deliver multiple TV channels in local environments. They use microwave antennas to deliver services directly from a head end to an antenna at the customer's home. An MMDS/MVDS transmitter will typically reach homes within a 15–50 km distance. All receiving homes must be in line-of-sight.

Another relevant technology is "LMDS" (Local Multipoint Distribution Service), which uses *two-way* microwave transmission. Transmitters for LMDS will normally reach homes within up to 2–5 km distance and will also require that all receivers are in line-of-sight.

4.7.4.1 DVB

While digital terrestrial most often is associated with broadcast of digital television to set top boxes, the most interesting use in the longer term may in fact be broadcast to mobile environments. ZDF, the national German broadcaster, has, for instance, in co-operation with Deutsche Telekom, Nokia and The Fantastic Corporation demonstrated successful DVB-T broadcast of multimedia at 14 Mbits/s to hand-held devices. Another project, involving Deutsche Telekom, BMW, the Fantastic Corporation and various content partners demonstrated continuos broadcast of multimedia to moving cars at 16 Mbits/s.

DVB is not any physical path of transportation of data; it is a project that has defined standards for digital TV distribution over different physical media such as satellite, cable, digital terrestrial (including mobile receivers), and microwave point-to-multipoint video distribution systems. DVB includes over 170 organisations from 21 countries, from all over the world. DVB uses MPEG-2 standard to distribute video, and as we have already mentioned, also data.

4.7.4.2 DVB/MHP

In 1997 the DVB project decided to expand its scope to include standard specifications for the hardware that received a DVB signal as well as for in-home retransmission of the content. This was called the "DVB Multimedia Home Platform".The work to develop this open standard commenced in 1997, and the first draft of it was approved in February 2000 and published in July 2000. HMP provides a technical solution for the user terminal that enables the reception and presentation of applications in a vendor, author and broadcaster "neutral" framework. It includes a set of APIs (Application Protocol Interfaces) based on java that enable any technology provider to develop applications for it. It can be used for applications such as:

- Appealing electronic programme guides

- Enhanced broadcasts (sports, shows,...)

- Interactive applications (web browsing, e-mails, distance learning,...)

- E-commerce

- Interactive games

- News services

The platform, which is now developing in promising ways, assumes several ways to navigate through broadcast content. Firstly, once the receiver is switched on, the consumer will see an overview over all services received. This is based on subscription information data. It identifies and allows selection of all available events to which the user could be entitled. In addition to this there is an electronic programme guide, which gives more detailed information about the services.

While DVB as a transport stream specification already is hugely successful, it will be beneficial for the media industry if also the extension – the Multimedia Home Platform – takes off as is now widely expected.

4.7.4.3 Digital Audio Broadcasting (DAB)

Radio broadcasting is traditionally based on FM or AM, which are analogue signals and thus not real "data broadcast" signals. FM (or AM) is what you get in the radio in your home or car. And it is what your parents got: FM was invented in the 1940s, AM in the 1920s. One of the things you often will have noticed when using these technologies is that the signal deteriorates considerably from time to time. This is caused by so-called "multipath interference" when the analogue signal bounces off buildings, trees and hills and arrives at your receiver out of phase with the main signal, confusing the transmission. This is a typical problem of analogue broadcast.

The alternative is digital broadcast. There exist several standards for digital audio broadcasting such as DSR/ADR (for satellite), but the most interesting is DAB. This is an open standard, which has been under development since 1981 at the German Institute für Rundfunktechnik (IRT) and since 1987 as part of a European research project (Eureka 147). The technology uses about 175–200 kbps to transmit one audio channel which sounds as good as a CD or DVD. It uses, in other words, 3–4 timers as much as what you can get through a standard 56 kbps modem, but a lot less than what you need to receive a single, digital TV signal.

The system is cheating a bit with the data it transmits. A CD actually delivers a signal of approx. 1.4 Mbps – about seven times as much as the DAB signal. However, the DAB compression has been based on stringent testing in the international standardisation procedure, including numerous listening tests involving hundreds of people. The respondents first hear the original and then two other versions of a spoken/musical item, of which one is coded and the other is the original version again (they do not know which is the original). Respondents then rank each of the two pieces on a scale from 1 to 5 (1 = very poor quality, 5 = indistinguishable from the original). The results are analysed statistically. These tests have consistently shown that for nearly all types of speech and music, at a data rate of 192 kbps and over, on a stereo channel, scarcely any difference between the original and coded version was observable (ranking of coded item >4.5).

DAB provides a number of advantages over FM or AM. Firstly, since it is based on digital data, it is designed in such a way that broadcasters can deliver text information about what you are hearing. They can also identify their channel by its name rather than frequency, and they can add descriptive information that enable the listener to search for channels with different characteristics. You can ask your DAB receiver to search for all the radio programmes, which match your request – like Top 40. It will find the programmes that have that identification. Broadcasters will have the option to transmit additional Programme-Associated Data (or PAD for short). This might include comprehensive information about a piece of music being played, such as the song title, composer, singer, album name and number and so on. The lyrics of an opera could be transmitted while it is in progress. And commercials could have accompanying text messages with more information about special offers, or contact phone numbers. Since DAB is digital, any information that can be digitised could be transmitted on DAB. Other services that might be implemented include text-based traffic and weather information, emergency warnings, paging services, financial services like stock market reports and exchange rate news.

4.7. Mobile

The largest market for mobile broadband multimedia will be within the area of "mobile phones". Mobile phones will go through a series of transitions during the coming years. A key distinction here is between the so-called "generations":

- First generation phones ("1G") were analogue phones with very limited functionality.

- Second generation phones ("2G") are digital phones with enhanced functionality.

- Third generation phones ("3G") are digital phones with broadband communication speed.

Most people use 2G today, and 1G is being phased out. The leading technology within 2G is GSM, which is used in all of Europe, much of Asia/Pacific and in selected areas in the USA. However, there is a migration path between 2G and 3G, where the intermediary solutions are called "2.5G" or "2 + G". 2.5G (as we shall call it) will comprise of three different technologies:

- HSCSD. This stands for "High Speed Circuit Switched Data". The technology uses circuit switching, which means that a dedicated connection is established between two people connecting. This is a direct continuation of the current 2G technologies and involves very limited investments and increases data transmission rates from the current 2G standard of 9.2 kbps (which feels very slow) to 57.6 kbps (which is standard modem connection speed for PCs today). The technology supports broadcast. However, it reduces the bandwidth available for traditional voice communication and most operators will probably skip this technology in their rollout.

- GPRS. This stands for "General Packet Radio Service". Package switching means that content is divided into digital packages that may share parts of the bandwidth even if they have different final destinations. The Internet is based on packet switching. GPRS is thus more bandwidth efficient and more future proof (thanks to its use of packet switching) than HSCSD. It supports data communication speeds of up to 115 kbps, which is twice the traditional modem speed. Furthermore, this technology supports broadcast. It is expected to be more widely deployed than HSCSD.

- EDGE. This stands for "Enhanced Data in the GSM Environment" and is a further development of GPRS providing more speed. Details are not yet fully known and commercial planning for take-up cannot be carried out until the release of more information.

After 2.5G comes of course 3G, which will use three different flavours of a technology called "CDMA" (which stands for "Code Division Multiple Access"). The different flavours of this are "W-CDMA", "TD-CDMA" and "CDMA2000", with the former being seen as the potentially most successful permutation. These different flavours arise because of regional legacy issues in the first and second generation systems and because of the bureaucratic/technical and political battles between the USA and the rest of the world in the area of standards. Basically the USA sphere of influence uses a mixture of standards, and the rest use GSM.

CDMA enables data rates of up to 2 Mbps, which is enough (more than enough, in fact) to support full motion video. However, this does not mean that all 3G phones will operate at 2 Mbps all the time. There are two reasons for this:

- Communication speed depends on the speed with which the user is moving around. While the communication speed may be 2 Mbps when the user is stationary it falls to 384 kbps when the user walks around with the phone and to 144 kbps when the user is driving a car. It is possible to broadcast good quality video to a limited size screen at 384 kbps, but at 144 kbps it will most likely need to buffer at the receiving device before playing out. This is more of an issue of signal quality and signal strength than speed. At speed, the issue of bandwidth is affected by the delays and processing in the network to affect a hand over, which brings a small break in the signal, which interrupts the flow of data. Hand over is dictated by the network conditions, and can also happen even when a user is not moving.

- Operators experience a trade-off in their choice of bandwidth offered. They can either offer the full bandwidth to all phones or limited bandwidth to a higher number of phones, or a combination (different service tiers).

The latter issue sounds a lot like the compromises that we experience on the Internet today. However, there may be two ways around this dilemma for data broadcast implementation. Both enable data broadcasting.

4.7.5.1 Data Broadcasting to Mobile Devices

The two approaches to simultaneous delivery of broadband media to mobile devices are very different. One uses a single networking approach while the other combines two approaches. The alternatives are:

- *Broadcast over the phone network.* Data broadcasting scales, as previously discussed, much better than point-to-point communication and a broadcasting approach will thus enable the delivery of broadband content to many without eating up too much bandwidth.

- *Broadcast in a by-pass of the phone network.* This option is to combine the dedicated phone transmission network with digital terrestrial broadcast technology – either DAB or DVB. This enables broadcast in a pure one-way stream to any mobile device. Such combination devices with two-way and one-way technologies have, for instance, already been successfully tested by Fantastic with Nokia, Deutsche Telekom and BMW.

As for broadcasting over the phone network, there are several technical solutions:

- Point-to-multipoint broadcasting

- Point-to-multipoint multicasting

- Groupcall

- Cell broadcasting

A special consideration here is the way the mobile phone operator structures its transmission network. A mobile operator will normally initially provide a thin lawyer of coverage over a large geographical area as illustrated in Figure 4.1.

Once the traffic goes up and the operator gains some experience with specific use, the operator will deploy a finer network of smaller cells, which will be allocated bandwidth in accordance with the average traffic (Figure 4.2).

The third step that could be taken is to add a new macro-cell for the whole area, which is allocated for data broadcasting of rich media such as video clips, etc. (Figure 4.3).

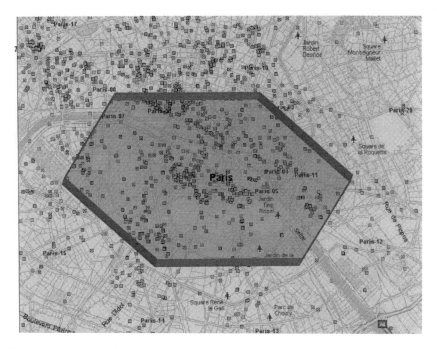

Figure 4.1 Mobile phone operators will often initially deploy a thin coverage from a single transmitter for a given area

Most operators choose a migration path towards 3G where they invest in 2.5G at an early stage (which costs in the tens of millions of dollars) and add an upgrade to 3G later on (3G can cost billions of dollars to implement for a carrier). Most of them will maintain their 2.5G technologies in operation in remote areas and only implement 3G in more dense population areas.

Figure 4.4 shows the two main migration paths. The upper migration path on the diagram (GSM-HSCSD-EDGE-3G) is interesting since it incorporates ability to broadcast, for instance, video clips in near-real time. However, the alternative migration path (GSM-GPRS-3G) will be more common. In any case, once 3G rollouts are reached there will be the possibility to deliver rich multimedia in broadcast and on demand.

Another issue that carriers will need to consider is which data services they will want to support. 2G (GSM) supports messaging, which requires very little bandwidth and which is asynchronous (content is downloaded when the phone is turned on, and it is not read until all of it has been received). Furthermore, 2G combined with WAP (Wireless Application

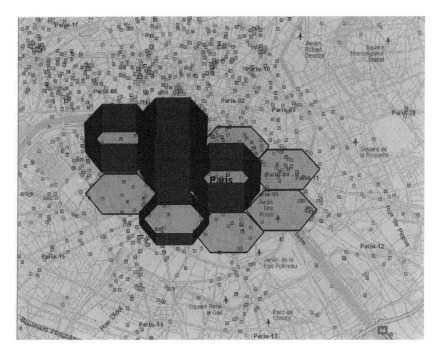

Figure 4.2   The network operator deploys additional cells as traffic grows in the area

Protocol) or DoCoMos I-mode supports the linking into specially designed web pages for access to text and figures.

GPRS supports more bursty data that requires a faster connection, but not data broadcast. HSCSD and 3G supports (with some limitations) broadcast and 3G even broadcast at broadband connection speeds. This technical landscape is illustrated in Figure 4.5.

One final aspect of broadband delivery to mobile applications is the cost of licenses. An interesting case: the British Government offered in 1998 the one and only nationwide DAB license for sale. The terms were attractive: 12 years exclusive license with automatic roll-over for another 12 years. The one and only bidder – GWR with NTL - paid £50,000 for the license. The government proceeded to auction its computing UMTS licenses the year after. The combined priced paid by the five winning bidders was £22,500,000,000 – 450,000 times as much. It might be argued that the relative valuations of these rights were not in full proportion to their potential commercial value.

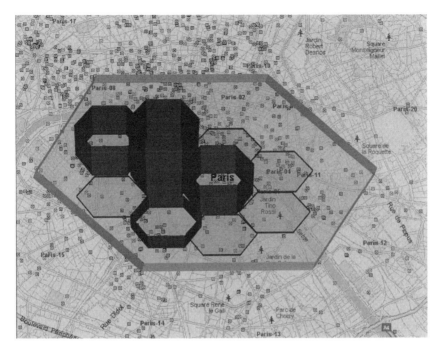

Figure 4.3 The network operator can add a macrocell for data broadcasting to minimise data congestion

4.7.6 Satellite Distribution

Although one of the "traditional" data distribution systems, satellite still offers very special advantages for some situations, not achievable by any other distribution medium. The most significant of these is the ability to reach geographically scattered audiences. There are three main concepts in satellite technology:

- *Geostationary Satellites (GEOs)*. These are located in the Clarke Ring (35,750 km above the Equator) and are pointed at a fixed part of the globe.

- *Medium Orbit Satellites (MEOs)*. These are orbiting at 6,000–12,000 km from the Earth.

- *Low Orbit Satellites (LEOs)*. These are orbiting at 200–3,000 km from the Earth.

Two very widespread concepts for satellite communication are:

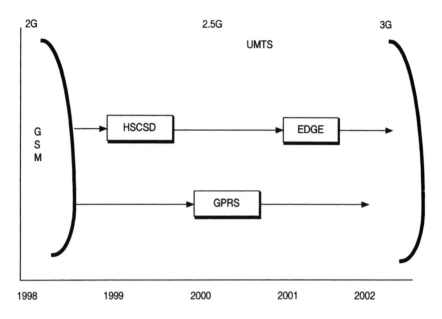

Figure 4.4   Typical migration paths from 2G towards 3G communications

- *Very Small Aperture Terminal (VSAT).* Smaller stations used for two-way communication with GEOs. They have typically been deployed for corporate solutions and use dishes with a diameter of 0.9–1.8 m. The typical transmission speed is about 64 kbps to 2 Mbps.

- *Direct broadcast satellite.* These are used for broadcast of television signals. Direct broadcast satellites are, unlike VSAT, used for mass-market applications. As the name indicates, the concept is that you broadcast directly to end-users, rather than to intermediary redistributors. Direct broadcast satellites are geostationary, meaning that they are orbiting at 35,780 km above the Equator, so that they are at a fixed point as seen from any given point on the Earth.

Since satellite networks are inherently broadcast, multicast through them is easily accomplished, as it is a subset of broadcast.

4.7.7 Home Networks

''Home Entertainment Networks'' is widely seen as one of the major new growth areas in telecommunication. The assumption is that people will

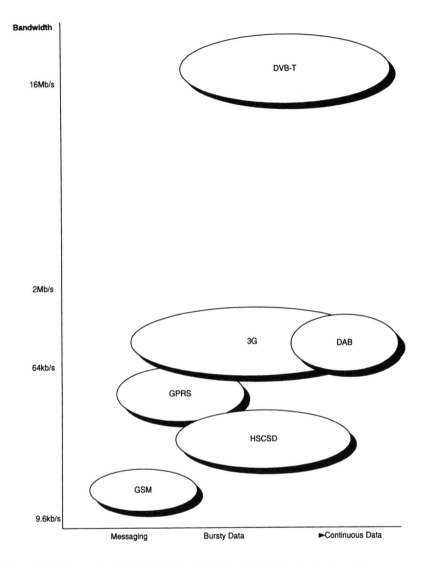

Figure 4.5 Comparing 2G, 2.5G, 3G and DAB and DVB-T technologies with respect to broadcasting support and bandwidth

have access to a wide range of IP devices, and that they will want to connect these devices to each other – and to flows of media from the outside. The growing use of mobile computers, such as notebooks and laptops, compounds the problem because users of these computers typically need to change their PC configurations on a more frequent basis. These mobile devices are often unplugged from corporate networks and peripheral devices in the office, and re-configured to enable computing and remote communication from the home or while on the road. A major benefit of home networking can be that digital content received from digital broadcast, the Internet, and package media can be distributed throughout the home to any appliances on the network. Interoperable digital AV appliances – used in conjunction with the appropriate digital content protection technology – will provide an infrastructure for home AV server applications, Internet communication, and for the last few metres of distribution for data broadcasting services. Combining DVB-HMPO with home networking makes for interesting applications. Broadcasting might deliver data to an MHP set top box with embedded hard disk – this data is then accessed via a home network thus providing a walled garden of readily available on-demand content.

Several technologies are emerging to design intelligence into the PC itself to handle installation and configurations without user intervention:

- *Bluetooth*, an open standard for narrowband wireless data transfers over short distances.

- *IEEE 802.11*, an open standard for broadband data transfers over short distances.

- *HomeRF*, an open standard for broadband transfer over short distances.

- *IrDA*, an open standard for broadband transfer over short distances.

These four home networking standards are briefly described below.

4.7.7.1 Bluetooth

Bluetooth was the name of a Danish Viking king, who was able to convince many wild tribes to work together (a rare skill, which is also in demand in electronic communication markets!). It is also the name of a short-distance wireless communication technology (set of technologies, in fact), which is poised to become the most important solution for home entertainment networks as well as for much of the inter-device communication in offices and elsewhere. The initiative was originated by Ericsson, Intel, IBM, Nokia and Toshiba, but more than 2,000 other manufacturers had by mid 2000 joined the family. Arguably, this makes it by far the fastest growing industry standard ever.

Any device that uses the Bluetooth standard for communication will contain a tiny microchip, incorporating a radio transceiver. This will enable the device to communicate without use of cables. The communication will facilitate fast and secure voice and data transmissions, even when the devices are not within line-of-sight. The radio operates in a globally available frequency band, ensuring compatibility world-wide.

The Bluetooth consortium talks about "personal area" communication – a term that builds on Wide Area Networks (WANs) and Local Area Networks (LANs). A Personal Area Network (PAN) based on Bluetooth will provide the following advantages:

- *Simplified access to external networks.* This is done by recognising and connecting to different types of networks through a Bluetooth connection. For instance, you can just as easily and instantly connect to the Internet via a mobile phone as via any Bluetooth-enabled wirebound connection.

- *Cable replacement.* The technology eliminates the need for troublesome cable attachments. You can, for example, send and receive e-mail on your mobile computer via your mobile phone, even when the devices are not within line-of-sight.

- *Personal ad-hoc networks.* All your Bluetooth-enabled devices can be set up so that they automatically exchange information and synchronise with one another. For instance, if you accept an appointment on your hand-held device, the appointment is automatically accepted in your desktop PC as soon as the two devices are within range of each other.

The Bluetooth technology is designed to be fully functional even in very noisy radio environments (however, as IEEG 802.11 and Bluetooth both operate in the unrestricted bandwidth area, they will in reality sometimes interfere – more about this later). The technology provides a very high transmission rate and all data is protected by advanced error-correction methods, as well as encryption and authentication routines for the user's privacy.

The following (provided by the consortium) are examples of what the technology will enable:

- *Mobile phones. Your gateway to the world.* In the Bluetooth world your mobile phone is automatically connected to other digital devices.

You could, for instance, connect your mobile computer to the Internet and send and receive e-mails, even when the linked devices are not within line-of-sight. The mobile phone becomes your gateway to the world.

- *Same phonebook everywhere.* For practical reasons, we tend to have one phonebook in our mobile phone, one in our mobile computer and/ or desktop computer at our office and one in our hand-held device. With the Bluetooth technology, all of these phonebooks can be instantly and automatically synchronised each time the devices come into range. This makes it possible, for instance, to key all phone-book updates into your computer and then easily and swiftly transfer them to your mobile phone – or vice versa. Same phone for all needs. By making your office intercom system Bluetooth-enabled and adding a Bluetooth-compatible base station at home; you can use the same phone wherever you are. At the office, your phone functions as an intercom, with no telephony charge. At home, it functions as a porta-ble phone, with a fixed line charge. And when you are on the move, the same phone functions as a mobile phone, with a cellular charge. The switches will be made automatically depending on which network is available within reach.

- *Mobile computers.* Your information warehouse. Most of the infor-mation you need on a day-to-day basis is stored in your mobile computer. Therefore, it is also your natural information warehouse. With the Bluetooth technology, all connections between your mobile computer and other Bluetooth-enabled devices are instant and auto-matic. You can send files from one mobile computer to another as easily as for example over a LAN. Or you can surf the Internet regard-less of your location – through a mobile phone or any Bluetooth-enabled wire-bound connection. Sending and receiving e-mail and faxes are just as simple. As long as your mobile computer is within reach of a Bluetooth access point, you have a fast, secure and wireless connection to the outside world.

- *Hand-held devices.* Your hand-held companion. All the information stored in your hand-held device will be accessible from other Blue-tooth-enabled devices as well – without connecting them by cable. And vice versa. Important records, such as your calendar, contact list, phonebook and to-do list, will never have to be outdated in any of your devices regardless of where and when the information is entered.

With your hand-held device wirelessly connected to a mobile phone, or any wire-bound Bluetooth-enabled connection, you can actually send and receive e-mail, notes and simple documents by tapping on the screen. The connection is established automatically and will be maintained even if the devices are not within line-of-sight.

- *Headsets.* Keep your hands free for more important tasks. The wireless headset is as easily connected to a mobile phone as to any other Bluetooth-enabled wire-bound connection. Through the headset, you can automatically answer incoming calls, initiate a voice-activated dial-up and end a call – if your mobile phone is equipped with proper functionality. You can also transfer a call or alter the ringing tone between the headset and your mobile phone or a mobile computer. The wireless headset offers very high quality sound and allows audio playback from a mobile computer. You can also control the volume and microphone with the headset from a computer or mobile phone

- *Office equipment.* Eliminates all troublesome and ugly cable attachments. The Bluetooth technology connects all office peripherals wirelessly. You can, for instance, connect your desktop or mobile computer to printers, scanners and faxes without ugly and troublesome cables. And you can increase your sense of freedom in your everyday work by a wireless connection of your mouse and keyboard to your computer.

- *Cameras, still image and video.* Lets you send instant postcards from any location. The possibility to transfer still images and video clips between a camera and a mobile computer is a good example of the versatility of the Bluetooth technology. When your digital camera is Bluetooth-enabled, you will be able to send instant postcards as still images or video clips from any location by wirelessly connecting your camera to your mobile phone or any wire-bound connection.

- *Other electronic devices.* Propels you into a new dimension of wireless connectivity. The potential of the Bluetooth technology is virtually unlimited. One after another, new applications and products, as well as increased functionality, will be introduced. From small hand-held scanners, portable hard disks and wrist watch information centres to refrigerators, coffee machines and presentation projectors are just a few examples where fast and secure wireless connection will simplify our everyday life. As for all other Bluetooth applications, all connections are instant and automatic, and maintained even if the devices are not within line-of-sight.

- *Internet access and e-mail*. Bring the world to you, wherever you are.
  The Bluetooth technology brings the world to you regardless of your
  location. From your desk-top computer, mobile computer and hand-
  held device you will always have fast and secure wireless access to the
  Internet – through a mobile phone (cellular) or through a wire-bound
  connection (PTSN, ISDN, LAN, xDSL). To connect is as simple as
  switching on lights.

- *Office LAN (Local Area Network)*. Redefines the meaning of flexibility. By installing a Bluetooth network at your office you will eliminate
  the troublesome and ugly cable attachments and redefine the meaning
  of connective flexibility. No longer will you be bound to certain
  locations for connections or have to install new cables for new work-
  stations. Each Bluetooth-enabled unit can be connected to more than
  200 other devices. And since the technology supports both point-to-
  point and point-to-multipoint connections, the maximum amount of
  simultaneously linked devices is virtually unlimited. Furthermore,
  whether you are working in mobile mode or back at the office, your
  mobile computer can automatically be connecting to the LAN.

4.7.7.2 IEEE 802.11

Bluetooth will co-exist with another open standard with a less compelling
name. It is called "IEEE 802.11". So what is this?

It can perhaps best be described as the wireless version of corporate
LANs. The dominant LAN approach is based on the so-called "Ethernet"
solution, which was developed by Robert Metcalf in 1970. Metcalf
learned in 1970 about a network called "AlohaNet", which was used
for data communication between the Hawaiian Islands. It worked in a
very simple way. Content was divided into small packets that each had
a digital header with information about where it should go to, and a
payload, which was the actual content – just like the Internet today. The
key feature of AlohaNet was that you did not have to "wait for a dial
tone" when you wanted to send something to someone. You could just
shoot off your message at any time, and each of the digital packages would
then go on its way towards the final user, and once it got there, the final
user's computer would send back acknowledgement information. If your
computer never received the acknowledgement information then it could
conclude that the digital package was "lost in the ether", as Metcalf
expressed it, and your computer would then just send it again. This

concept is today called "package switching", as opposed to the dial-tone concept, which is called circuit switching".

The leading open standard for Ethernet is called "802.3" and was defined by IEEE ("I-triple-E"), which stands for "The Institute of Electrical and Electronics Engineers, Inc.". This organisation describes itself as follows: "helps advance global prosperity by promoting the engineering process of creating, developing, integrating, sharing, and applying knowledge about electrical and information technologies and sciences for the benefit of humanity and the profession." Early generations of Ethernet (and much of what we have in offices today) provided 10 Mbps transfer rates, and most modern office networks provide approx. 100 Mbps, and currently moving towards the gigabit range.

The development of Ethernet standards took a turn after a number of companies formed around 1991 began to make wireless Ethernets. These would use unlicensed frequencies in the transmission spectrum to deliver niche applications such as communication with portable hospital computers, and to reduce the cost of wiring schools and universities. IEEE began consequently to develop an open standard for high-speed wireless networks. The first specification was released in 1997, and the name was "802.11". IEEE 802.11 standard supports transmission in infrared light and two types of radio transmission within the unlicensed 2.4 GHz frequency band:

- Frequency Hopping Spread Spectrum (FHSS)

- Direct Sequence Spread Spectrum (DSSS).

4.7.7.3 HomeRF

HomeRF has been developed by the "HomeRF Working Group", which has representatives from a number of companies in the industry. It works with higher bandwidth and transmission distance than Bluetooth, and can enable applications like:

- Setting up a wireless home network to share voice and data between PCs, peripherals, PC-enhanced cordless phones, and new devices such as portable, remote display pads.

- Access the Internet from anywhere in and around the home from portable display devices.

- Share an ISP connection between PCs and other new devices.

- Share files/modems/printers in multi-PC homes.

- Intelligently forward incoming telephone calls to multiple cordless handsets, FAX machines and voice mailboxes.

- Review incoming voice, FAX and e-mail messages from a small PC-enhanced cordless telephone handset.

- Activate other home electronic systems by simply speaking a command into a PC-enhanced cordless handset.

- Multi-player games and/or toys based on PC or Internet resources.

The standard uses the so-called "Shared Wireless Access Protocol".

4.7.7.4 IrDA

IrDA stands for "The Infrared Data Association", an organisation founded in 1993. The organisation was inspired by the infrared remote controls we use for television and other devices when it started developing its standards. These have been widely used for the infrared links in note-book PCs and elsewhere. The standard normally assumes just two devices talking together, and over very short distances (up to 1 m). It can provide very high bandwidth for mobile phones and PDAs, though – up to 16 Mbps.

4.7.8. Comparing Wireless Local Networking Standards

Table 4.4 provides a simple overview of the four open standards for home entertainment networks (and in two cases, wireless LAN).

Bluetooth is a narrow-band technology (although much faster than a 56 kbps modem), and it has a short rage. The advantages are the very low power consumption and hardware prices that go with the limited perfor-mance. However, the low power consumption can also be a problem, since for instance IEEE 802.11 with a much stronger signal can interfere with Bluetooth if they are both used on the same location.

Table 4.4 A simple overview of the four open standards for home entertainment networks

	Bluetooth	**IEEE 802.11**	**HomeRF**	**IrDA**
Commercial positioning	Mainly consumer, narrow-band for very short distance	Mainly corporate broadband	Mainly consumer, for short and longer distance, narrowband or broadband	Consumer and corporate for very short distance, narrowband or broadband
Speed	30–750 kbps	11 Mbps (higher bandwidth available soon)	1–10 Mbps	9600 kbps to 16 Mbps
Range (m)	10	15–100	50	1
Types of terminals	Built into notebook, cell phone, palm device, pager, appliances, car	Add-on to notebook, desktop PC, palm device, internet gateway	Add-on to notebook, desktop PC, modem, phone, mobile device, internet gateway	Build into notebook, desktop, printers, phones, pagers, cameras, watches, medical and industrial equipment
Typical configuration	Point-to-point or multiple devices per access point	Multiple clients per access point	Point-to-point or multiple devices per access point	Point-to-point

4.8. Reaching the end-user

We have now looked at electronic packages with content that is passing through different networks on their way to a final destination. But what is that destination?

Traditional mail is delivered to houses, flats, office buildings and mansions. The final destination for Internet content is today almost entirely PCs. However, there will soon be a multitude of other devices that can receive digital content. The IP anywhere era started with PCs, but the next devices will be set top boxes (as digital television is rolled out), and mobile devices (especially with the commercial rollout of 3G technology, which will enable broadband), and finally a multitude of household utilities. This provides a key challenge for the media industry; since content often will flow through numerous different topologies from the time it leaves the content provider until it reaches the final user. The complexity is growing considerably.

5. The Five Basic Formats of Data Broadcasting

"I have travelled the length and breadth of this country and talked with the best people, and I can assure you that data processing is a fad that won't last out the year."

The editor in charge of business books for Prentice Hall, 1957

We have now looked at the surrounding technologies that we will encounter when we want to launch a data broadcasting service. We looked at existing networking infrastructures and technologies, and specifically:

- How you divide content into packages

- How you choose delivery type

- How you specify destination

- How you provide guaranteed unique addresses

- How you provide information about how to reach destinations

- How you choose a vehicle for transportation

- How you choose a physical path

- How you reach alternative end-user device

Chapter 6 will be devoted to the task of building a complete data broadcasting platform, and following that we will move on to look at how you run services and applications over such a platform (Chapter 7). However, before we get to those stages we should define what kinds of content from a technical point of view you would consider to run over a data broadcasting platform.

5.1 THREE KEY QUESTIONS REGARDING THE DATA BROADCASTING CONTENT

Different people will think of different services when they hear about "IP multicasting" or "data broadcast". One may think of it as a way of broadcasting web pages, another as a vehicle for interactive television, a tool for providing real-time financial data or as a solution for distribution of software packages. However, each of these concepts would require a rather different broadcasting approach, and data broadcasting encompasses all of these approaches. So let's look at the basic criteria that distinguish choice of basic technology questions. The three most important questions are:

- Is guaranteed delivery necessary?

- Is the delivery time-critical?

- Is content viewed on delivery?

5.1.1 "Guaranteed Delivery" or "Best Effort"?

"Guaranteed delivery" is the concept that we know from registered mail. Broadcast data delivery using IP multicast is in fact usually not guaranteed. It is, instead, what you call "best effort delivery".

Best effort delivery has the advantage of working without interruptions and without requiring acknowledgements via return path. This makes it very fast. Furthermore, there are many media concepts where occasional loss of data is not a prohibitive problem. Two examples:

- *Web based content.* The content of a website will usually consist of many different files, and it normally will still be of some use even if one or two files are missing.

- *Audio/video streams.* Here you may hardly notice if a few frames are missing. This is often the fact in satellite television transmission, when weather or structural interference disturbs it (still pictures of the last frame are usually broadcast during the feed interruption to cover for missing live frames).

However, assume that we are not sending web pages or streaming audio/video. No, we want to broadcast *software updates*. With software updates you can normally not afford to lose even a single bit in a file. Where completeness is a requirement a guarantee delivery mechanism is needed. (We discussed reliable multicast file-transfer in Chapter 4.)

5.1.2 Time Critical Delivery?

There are many applications in the market where it is vital that the content is delivered in "real-time". An example is professional stock market information. When a bank chooses a supplier of an information system for their dealers they will often ask for a free trial period when they put the competing screen systems next to each other. They will then track how quickly changes in financial price quotes appear on each of the competitive systems. A good system might update within 0.4 s from the time that the price change took place, while a less competitive system could take 0.6 s to get the update in the market prices through to the dealer. The dealer with the slower system will on average quote prices that are behind the market, and the 0.2 s can cost millions of dollars in lost revenues over a year. A unicasting based Internet system, which has no Quality of Service (QoS), would obviously not even be considered.

5.1.3 Viewing on Delivery?

Radio and television is basically based on viewing on delivery. This concept has two advantages:

- You do not need to store the content locally (saves time and storage space).

- Content can be displayed in real-time.

The down-side of view-on-delivery is of course that you cannot view after delivery unless the solution is combined with a storage option.

5.2 DATA BROADCASTING FORMATS – AN OVERVIEW

We have now looked at three basic criteria for selecting basic data broadcasting format: "guaranteed delivery", "time critical content" and "viewing on delivery". But what are the formats you can select between? There are five alternatives:

- Cached content delivery

- Audio/video streaming

- Package delivery

- Real-time data streaming

- Broadcast guide data

We can summarise these basic delivery formats and characteristics in Table 5.1.

Table 5.1   Overview of data broadcast formats

	Cached content	**A/V streaming**	**Package delivery**	**Real time data streaming**	**Broadcast guide data**
Guaranteed?	No	No	Yes	No	No
Time-critical?	No	Yes	No	Yes	Yes
Viewed on delivery?	No	Yes	No	Yes	No

We will in the following section take a closer look at each of these five basic formats.

5.3 CACHED CONTENT DELIVERY

Content that is delivered to the end-user and stored on the hard disk is called "cached content". Cached content delivery is best effort delivery, as the content provider and the broadcaster have no means of knowing if the content was successfully received by the receiver. Access to the cache is typically provided through an Internet browser, and the cache is automatically managed by the receiving system. This format supports content created with standard web authoring tools.

Below we will describe mainly the features and issues related to receiving and deleting cached content. Section 5.3.2 will describe the scheduling and broadcasting of cached content.

5.3.1 Receiving and Deleting Cached Content

5.3.1.1 Advantages of Cached Content Delivery

The concept of cached content provides five major advantages:

- *Immediate availability for viewing.* Caching solutions are often provided to accelerate access to information. Cached content delivery in a data broadcasting system delivers the data directly to the end-user, therefore creating a local cache. Access to the cache is as fast as access to the local hard disk.

- *Broadcast and view times can be different.* A disadvantage of stream based broadcasts like television is that the user needs to view the delivery channel when the programme is broadcast and not when the user is ready to view the programme. The television business invented the VCR to solve this problem. The VCR is, as we know, a mechanical device with some automated features that most users do not know how to handle. Cached content delivery solves this problem automatically for data broadcasting. A content provider schedules data for delivery at planned times. The receiving device accepts data on subscribed channels and stores the data in a local cache. The cache update can happen during off-hours or while the user works on other tasks on his PC or watches other programmes on a TV/set top box. At leisure, the user accesses the media through a web browser.

- *Supports multiple viewing streams.* Unlike data delivered and viewed sequentially in one stream, cached content consists usually of many different files, tied together by a set of HTML pages. A broadband distribution system allows delivery not just of text and pictures, but also high quality video clips, DVD quality/surround sound music/voice and background material contributing to one topic. This makes this delivery service especially appropriate for large volumes of data used for progressive viewing, discovery and exploration. When a user subscribes to a newspaper, it is unlikely that the user reads every single word in the paper. We find the same behaviour when users browse through cached content, but unlike a newspaper, the user has immediate access to large volumes of supporting and related material.

- *Suitable content widely available.* Web based content re-used in the broadcast environment or newly created web content is suitable for this delivery type. The nature of the system allows for frequent, inexpensive updates of content, showing clear superiority to the static nature of content distributed on CD-ROM.

- *Smooth link to the Internet.* Content providers can easily embed links to the Internet within the cached content (for situations where they can assume that the users have access via a return path).

5.3.1.2 How Does the Cache Work?

Figure 5.1 provides an example of how a cache may work in a data broadcast system. The relevant process are highlighted with fat arrows.

The key steps in the caching process on the client device are:

1 The module filters filter the data packets for valid subscriptions.
2 The filters then pass subscribed content packets to the dispatch.

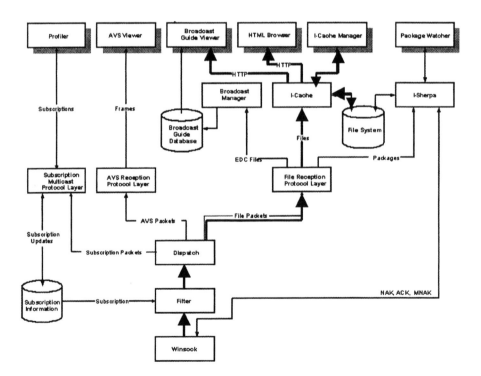

Figure 5.1 Example of a cache software topology for data broadcast

3   The dispatch will subsequently detect cached content data packets, and send these individual packets to the file broadcast reception layer.

4   The packages are assembled at the broadcast reception layer to create a complete file from individual packets. The file reception layer also can assemble files from out of sequence packets or can fill in missing packets from repetitive transmission of the same file or the same packet.

5   Each completely assembled file is sent to the intelligent cache handler.

6   The intelligent cache handler creates a hash entry for fast access from the content URL, stores the URL to file-name handling in a database and the file itself in the file system.

7   The content can be accessed from a browser or Broadcast Guide Viewer.

5.3.1.3 Viewing Cached Data

Internet browsers are the obvious choice to be used as viewers for cached content, since they are ubiquitous and designed to be extensible through scripts and plug-ins (data broadcasting can also be used to deliver these browser extensions to the client).

But Internet browsers were originally created as tools for finding contents on the Internet. How, then, will you know when to search on your own hard-drive (your broadcast content cache) and when to search on the Internet? The answer to that is that you do not need to know where the content comes from. The cache server installs itself as a local Internet proxy server between the Internet browser and the cache. Every HTTP request from the browser is thereafter intercepted by the proxy server and the content. If data matching a URL is found in the cache the content is returned back to the viewer, which means that normal Internet access, if present, is only used to retrieve information not found in the cache. This, by the way, raises two issues:

• *Considering external links in broadcast web pages.* You can broadcast parts of web pages to a user who you know is connected to the Internet. When he clicks on any link not pointing to content in his cache, but to content residing somewhere on the Internet, then he will still get it – just slower. But if you broadcast the same web page to a user without Internet connection, then he will simply get an error message when clicking on such a link.

- *Choice of URLs.* Names of URLs are (as previously mentioned) co-ordinated world-wide by various organisations. You apply for a domain name (a URL), and they will link it to a unique number. If someone else tries to set up a server with the same URL, then they will simply find that no one is able to link to it – the system is designed to prevent conflicts. However, data broadcasting is using URLs without requiring Internet routing to access them. This means that you could in principle define a URL that someone else has registered on the Internet, broadcast content that will be called up on the I-cache under that URL – and this content will then be what you see as the local search on the hard drive takes place before any contact to the Internet. The obvious way of avoiding such conflicts is to use domain names that are not recognised by the DNS system.

5.3.1.4 Cache Management – "the Digital Garbage Collection"

Data broadcasting is a fast and continuous process – a constant and often massive data stream is sent over the distribution media. If not managed a cache just continues to grow until the pre-set limit is reached or the memory fills up. So automatic solutions for cache clean-up – for digital garbage collection – are clearly needed. And as cached content delivery is a best-effort delivery only, caches need to be managed for consistency. There are several specific issues to consider here:

1 *Self-contained solution.* Cached content data management has to be fully self-contained. This requirement is imposed, as we do not require a path back to the content provider. We can put this in another way: the garbage collector has to come by himself, as there is no way of calling him.

2 *Data consistency.* To make sure that all components of a content set are delivered, the distribution mechanism has to impose a transaction model where related content is either fully delivered or not delivered at all. If we do not observe these rules, we might end up with inconsistent or incomplete content. An example could be a multimedia news channel where headlines, video clips, pictures and text were not matched on logical containers. A headline lost in transmission might then mean that all subsequent headlines were matched with the wrong stories. Alternatively we might see a headline, but when we click on it the story does not appear.

3 *Intelligent caching.* Traditional Internet technology uses a very simple digital garbage collection. The deletion of cached content is triggered

by reaching a pre-set cache size limit. It is here assumed that the end-user is the best cache manager (knows when to call the garbage collector), as he knows what he wants and does not need any more. However, the actual reality is that most end-users do not bother to manage this garbage collection, which means that they just get a first-in-first-out clean-up process. While this may work reasonably well for most traditional Internet users it will not be satisfactory for data broadcasting services, where you have massive amounts of content updating all the time (a user of a data broadcast service might receive thousands of times as much content as traditional Internet users). Assume, for instance, that you receive video clips as part of a music video channel. You would not want to store more than a few, since they are huge. But if you receive financial price quotes or text based news stories then there would be no problem in storing thousands of entries. Either the content provider or the user should consequently be allowed to manually specify a caching rule for each category of content.

5.3.1.5 Different Approaches to "Intelligent Caching"

Now, assuming that you want to pre-determine different garbage collection schemes for each part of the cached content within each channel or service that you broadcast. How do you do that? Firstly you should realise two basics about caching policies:

- They are always based on a "retain list". The cache always assumes content received, so the policy defines how it should be retained (if at all). Not being retained means of course that it is being deleted.

- The browsers are set to do some simple automatic retaining as a part of their caching policies. Intelligent caching needs to overwrite those policies while not jeopardising what happens at traditional Internet sessions.

Bearing these rules in mind we can now define different ways of retaining content that can be made to overwrite the browser's pre-set simple caching rules for data broadcasting sessions. Table 5.2 lists the alternatives.

One of the last options in the table – user-preference based garbage collection – has wide implications, since subscription information embedded in the data stream can provide enough granularity to have thousands (or billions) of channels. On television, we might have a sports channel, or even a sports channel specialised in automobile racing. On a data broadcast system, we might go further – we might, for instance, see channels with content related to stock car racing on concrete tracks in Arizona. Or we might choose to track one out of 220,000 financial instru-

Table 5.2   Overview of intelligent caching approaches

Intelligent caching solution	Relevant content examples
Since cached content is identified by URLs, the system should allow specifying a list of complete URLs. It may, for instance, specify that "www.re" should mean "www.reuters.com"	Any content
The system could specify URL patterns. It could, for instance, be that www.fantastic.com/ press* refers to all URLs below press*. Other standards than may be used are "?:", which means single character and "[a-z]:" which refers to a range	News headlines, news text, news pictures and news videos are sent out individually. However, it is vital that headline, text, picture and video about a single story are combined, viewed and saved in a consistent way on the client device. This can be done by setting rules that associate specific URLs or URL patterns with each other
Specific content can be discarded if items age beyond a certain time	This could, for instance, be old weather forecasts
Specific content can be deleted if it has not been used for a given period of time, or statistically not very often over time	This may, for instance, be one out of several alternative categories of advertisements
Specific content can be deleted when it has been used just once, or twice, or any other given number of times	An example would be an advertisement, where the advertiser wants the user to see different versions each time he clicks

Table 5.2 (*continued*)

Intelligent caching solution	Relevant content examples
Specific content can be deleted when a garbage collection hint is sent with the content. These tags can be broadcast as part of HTML pages and can be used by the garbage collection mechanism to make intelligent decisions	A provider of stock prices may, for instance, send an instruction by the end of each trading day to delete intra-day price movements and only store daily high, low, and close prices
The user can have the opportunity to set his own rules for continuous garbage collection for each item	A user receives streams of stock market information. At one point he decides to track tic-by-tic updates on a given stock in a chart. He then changes the settings for that item so that tic-by-tic data are stored for a pre-defined period of time, like, say, 10 days
The user can have the opportunity to set his own rules for *ad hoc* garbage collection for each item	The user can choose access to specific events, like live concert broadcasts where some of the video may or may not be cached
An agent on the user's receive device can determine how garbage collection should take place	An agent might recognise that a user is subscribed to the car-racing channel and detect that the local convention hall advertises a car show exhibiting rare sports cars. The agent alerts the end-user and displays information on the event. If the local news channel broadcasts a daily brief on the car exhibit, the agent subscribes to that channel on behalf of the user. Or, if a user subscribes to an out-of-town newspaper, he/she might not be interested in the daily events section. A usage tracking system could detect this and provide this feedback to the caching agent

Table 5.2 (*continued*)

Intelligent caching solution	Relevant content examples
Localised frameworks and local content insertion	If a content provider uses a standard template for content, it is a waste of bandwidth if the template is re-broadcast with the content. Consider an example of a news channel that has sections with business, sports, weather and entertainment news. What you want to achieve is that only the headline and picture changes for each of the sections, but not the template. This means that content is inserted into a local framework. Both are cached

ments in a finance channel in a different way than all the other instruments.

5.3.1.6 Content Tags

HTML supports a concept called a "tag". A tag is a code that identifies an element of a document. Some change text appearance, others let text or graphics act as a reference to another page, and still others define how the page will look. META tags are tags that define concepts in a page. META tags are, for example, used to identify a page's author, the version of HTML for which the page was written, and even keywords to help tell others what the page is about. These META tags are used by the receiver to make intelligent decisions about the content.

Content providers can use META tags to rate content. They can determine what age groups content is suitable for, warning about adult content and rating content for violence. Other standard tags can also be defined to determine the language in which content is written or spoken, and the country and region from which it originates or for which it is destined.

Content tags are used by HTML viewers to provide implicit content filtering.

Agents in data broadcast environments

Agents are small programs that operate on a user's behalf, performing a specific function. Agents running in the background do elaborate filtering based on user preferences. Agents establish content-related connections between information on different channels and present the user with these results.

5.3.1.7 Local Content Insertion in Intelligent Caching Systems

A typical intelligent caching system will support local content insertion. This may, for instance, be used for local ad insertion, where a large number of advertisements are sent to the local cache during low usage hours. These ads are subsequently automatically inserted when the surrounding broadcast content is received later.

The inspiration for such a solution can be found on the Internet, where web servers often insert local ads based on the overall content a user is requesting. A search engine might, for instance, insert advertisements for a DVD player into the result pages of a web-search on "DVD players" (a content provider/advertiser may buy the exclusive rights to launch ads when a specific search term has been entered into by the user). Intelligent agents and local content insertion can work together to replicate this service on the client in a data broadcasting environment. The advertisements here might not change as frequently as, for example, the news content that surrounds them. If a user is viewing a page describing a tropical destination the intelligent agent might choose to display an advertisement for an airline.

5.3.2 Scheduling and Broadcasting of Cached Content

5.3.2.1 Scheduling and Booking for Cached Content Delivery

Scheduling of transmission of cached content raises three general issues:

- *Facilitating the scheduling of complete websites or directories.* If we broadcast web content into a cache, then we need to be able to point the scheduling tool at existing web content. Now, scheduling this type of content file by file would be rather time consuming, so we have to look for alternatives. One such alternative is to use a "web-crawler" to copy a whole website into a local directory structure (web-crawlers are software tools that copy HTML documents, analyse the content and point the crawler at any links embedded in a page). The scheduler will in such a case analyse the website or a directory structure that you want to broadcast for completeness and consistency. The content provider needs in this case to explicitly allow sending web content that is not self-contained.

- *Flexibility in the scheduling time.* As cached content is not time sensitive we can give the booking system some flexibility in scheduling the exact start of the transmission and even in pre-empting the broadcast. A broadcast network might provide a lower bandwidth pricing for content with such a scheduling requirement

- *Using repetitive scheduling.* The content provider should have the ability to specify the number of repetitions per file as well as the number of repetitions per packet. The broadcaster will in that case use these parameters while broadcasting content. The content provider might use policies depending on content type to try to increase the probability of reception. An example: when sending a web page with content and advertisements the content provider might place more importance on the content than on the advertisement – obviously the advertiser paying for the ad might set different priorities. Higher repetition frequency might be set depending on importance of content. The user interface for a scheduling tool will automatically calculate the broadcast time based on data repetition parameters. The content provider will have to balance the cost of multiple transmission against the risk of losing some content.

5.3.2.2 Broadcast of Cached Content

As discussed earlier the cached content is not time-sensitive and does not require guaranteed delivery. The broadcaster has therefore some flexibility when broadcasting cached content. When bandwidth becomes scarce, the broadcaster might throttle back on the speed at which the cached content is sent out.

We have seen that the content provider can use data repetition to increase the likelihood of receiving all of the content. The broadcaster might use similar methods using available bandwidth. As most content in a best effort scenario will be fully received at the first try, the repetitive broad-casts are of less importance than the first one. A data broadcasting system can detect available bandwidth and use it to re-broadcast files. To do this, the broadcasting system reserves some space on the broadcaster's content storage for content with higher probability of incomplete transmission. Three factors influence this probability:

- The packet repetition factor

- The file repetition factor

- The size of the file

Large files that are broadcast without packet repetition and without file repetition are obviously more likely to be incomplete. The broadcast system can keep these files for a while in a *data carousel*, which is a tool that provides a revolving broadcasting of the same content. An agent will normally here monitor the available broadcasting bandwidth and insert packets from the carousel into the broadcast stream whenever band-width is available for it.

5.4 PACKAGE DELIVERY

Imagine that you are a software vendor. Package delivery is a method for guaranteed time-sensitive down-load of software that is installed on the user's device for instant or later use. You have just released new software, and you want to get it out to your customers immediately – or at least to those of them who are interested in buying it. So you decide to start a campaign. Everyone who has their receiving device turned on during the first 5 min of any hour will receive the whole package straight away. And furthermore, those who buy it will later be entitled to all kinds of upgrades and bug fixes that will be broadcast directly to their devices as soon as they exist. So which kind of broadcast format do you need in order do that?

Well, if you absolutely need to make sure that every single bit of your content is received, then you need to use the basic format called ''Package delivery''.

Package delivery is the digital broadcasting equivalent to registered mail. It is reliable, guaranteed delivery of content. Package delivery needs special considerations from the content provider, broadcaster and recei-ver. The section on reliable multicast file transfer discusses the technical issues involved in implementing package delivery, so in this section we will focus our attention more on the overall considerations and applications for reliable file-transfer.

5.4.1 Applications, Usage Tracking and Billing of Package Delivery

5.4.1.1 Applications

Package delivery can be used to deliver single files reliably as well as for packages containing multiple files bundled into one digital container. Package delivery can mainly be used for:

- *Software delivery.* Many software applications are frequently updated. Data definitions for virus scan tools, for example, are updated nearly weekly. Data broadcasting provides the optimal way to deliver updates at regular intervals, but it also allows the sending out of updates to many users at very little cost if a new critical problem happens.

- *Delivery of bought media.* Package delivery is also the delivery method of choice when delivering content, which the user has paid for individually. If a user buys music content on-line and creates a CD from delivered content, he/she wants to have a guarantee that all tracks are sent completely.

- *Magazine subscriptions.* Data broadcasting can be used to distribute magazine subscriptions. Magazines with multimedia applications are the best candidates for this type of delivery. Do not think of multimedia content as being related just to entertainment. Multimedia content for interactive display of trend data in a financial publication, video-clips from a political speech or daily reviews of the stock market might be other multimedia applications.

- *Business critical data.* A company using data broadcasting might send out financial spreadsheets, product overviews, product presentations and employee phone lists to remote offices. It is critical that data of this nature arrives complete.

5.4.2 Usage Tracking

We have already discussed usage tracking in the context of cached content delivery. However, when you use package delivery, you add additional requirements to usage tracking. So why is that?

The reason is that a content provider who uses package delivery is delivering content of particularly high value. It might, for instance, be a piece of software that cost $1,000. So the content provider wants to make absolutely sure that the content arrives, as the receiver will only pay for individual content (the $1,000!) if the content is delivered intact. Furthermore, you need to avoid the situation in which content is in fact delivered correctly but the receiver refuses to pay because he claims that it was not. Finally, you may want to ensure that the packages that you have transmitted are not just copied at will and used beyond the intended usage. This brings us to the concept of ''Object Based Billing''.

5.4.2.. Object Based Billing Models

Several object usage models might be supported by the data broadcasting systems with package delivery:

- *Buy.* This is the simplest of the revenue models for a package delivery based data broadcasting system. A user is here required to buy an object before being allowed to use it.

- *Pay-per-view/usage.* Pay-per-view is usually associated with video content, but it could also apply to games or other multimedia experiences, and then it becomes a *pay-per-usage* model. An example is that financial reviews of companies are delivered free, but each time a report is opened by the user he pays a certain fee. A game software vendor might associate a charge every time a game is played. Pay per view/usage for data content requires a system that supports micro-billing. A game software vendor might charge for every lap a user races around a race-track. Each lap might cost only fractions of a penny.

- *Rent to own.* The end-user never owns an object in a pay-per-view model. Rent to own is a combination of pay-per-view and buy. After a certain number of paid views the end-user owns the object. He has bought it through multiple usage.

- *Try-before-you-buy.* Try-before-you-buy (or ''Try and Buy'') is another variant of the buy model. A user can try an object for a predefined time free. If the user does not buy the object after the trial period is over, the object becomes unusable and might even self-destruct. This model is most usable for software distribution where trial use is a common feature.

- *Limited preview.* The try-before-you-buy model usually offers full functionality of a product. A limited preview model can work in multiple ways: short teaser of content, reduced quality content or limited functionality content. The first two methods are usually used for multimedia content, the last one for software distribution.

Both software and hardware based technologies exist today, which associate billing and revenue models with digital content.

5.4.2.2 Billing Model Technologies for Package Delivery Models

Billing can be handled with software only, or through hardware solutions. Software based technologies offer the advantage that they are available on many receiver platforms, while hardware based technologies offer more security. Software as well as hardware based technologies support two basic models:

- *Pre-pay model.* The digital money is loaded into a digital wallet on the receiver device. Each revenue transaction is stored locally and the transaction amount is deducted from the amount in the wallet. Vendors of billing solutions usually offer their services to refill the wallet. A log with all transactions is sent to the Service Company, which acts as a clearinghouse with the vendors. A wallet based approach is very well suited for micro transactions and for transactions where, for instance, a parent wants to control how much money their teenage children spend. Another pre-pay model is one where transactions within a given credit limit are approved, as we know from credit card payments.

- *Post-pay models.* These models are less common, as they are a virtual product/service advance against a promise of future payment. Models where each transaction is immediately executed are common for large transaction amounts, but are less suited for micro transactions, where the transaction cost might exceed the amount of the transaction.

Smartcard based systems also provide solutions to transaction billing. Their advantage is that they are more secure than software-only solutions. The Smartcard can act both as a wallet and as temporary storage for the transaction log. Smartcard based solutions have the additional advantage that the wallet becomes portable and personal. An end-user can use the Smartcard at other receivers and multiple end-users can share a receiving device and use their personal Smartcard.

5.4.2.3 Package Reception

A package receiver uses many of the same components as the best effort file receiver as we discussed under ''cached content delivery''. The package receiver can, like the best effort receiver, assemble files from out-of-sequence packages and assemble packages from repetitive transmission of packets. In contrast to the best effort receiver, however the package receiver does not wait for a repetitive transmission of the same file. This means that the package receiver has to be able to interact with the broadcaster through a back channel.

5.4.2.4 Package Delivery and the Back Channel

Package delivery assumed a broadcast connection for delivery and a back channel for confirmation. This does not necessarily mean that you must have a two-way broadband connection. You may very well use an existing Internet connection as return-path – or it might be a dial-up connection using a toll-free telephone number providing dedicated package delivery services. But how does the communication over this back channel work?

The process starts at the end of a package transmission session. The package receiver will then create a list with the missing packet numbers (if any) to ask the broadcaster for repeat delivery of the missing packets through the normal broadcast infrastructure, or using TCP/IP via the return path. The end-user should not have to care about that process. For the end-user the conversation of the receiver with the broadcaster has to be as transparent as possible. Only under extreme circumstances should the end-user ever know about problems with package reception.

5.4.2.5 Reliability of Package Delivery

In the discussion on cached content delivery, we define the probability of incomplete transmission based on:

- The packet repetition factor

- The file repetition factor

- The size of the file

The same three factors influence the probability of incomplete reception in a package delivery system. Each receiver who has not completely received a package will potentially ask for retransmission of missing packets. In a broadcast system, the complexity of a reliable delivery service grows rapidly with the number of potential receivers. In fact, because of the inherent nature of broadcasting, handling retransmission requests or reception notifications from a large number of receivers is directly dependent on the number of receivers. In other words, since in a broadcasting environment total reliability might not be technically achievable or financially acceptable, the term ''reliability'' has to be carefully defined. In that sense, we could define it as ''the knowledge on the result of the reception''. In other words, a transmission is considered reliable if the transmitter obtains the desired information on the quality of the reception, i.e. which receiver end-users have or have not successfully received the package.

The package delivery service should be designed to offer multiple levels of reliability. For this reason, different receipt notification policies are implemented. However, before we move on to the details, we need to define three alternative categories of notification:

ACTC: confirmation that the transmission went well
NATC: automatic notification that something went wrong
MNAK: manual notification that something went wrong.

5.4.2.6 Alternative Notification Receipt Policies for Package Delivery

5.4.2.6.1 Final Positive Acknowledgement Approach. In this pessimistic approach, we assume that *no news is bad news*. The receiver sends only a positive acknowledgement message (ACK) via the back channel to notify the broadcaster of his successful reception. An ACK also is sent if the receiver is subscribed (entitled) to the broadcast, but has chosen to turn off reception. Unsuccessful receptions are not notified. Within this approach, missing ACKs are interpreted as unsuccessful transmissions.

The advantage here is the reduced requirements for acknowledgement information: the disadvantage of this pessimistic approach is that the broadcaster has to keep a list of all subscribed receivers for the package. If a receiver is off-line it will never send an acknowledgement. The broadcaster interprets this as *bad news* and re-broadcasts the package until the file re-broadcast parameter as set by the content provider is exhausted. With a non-working receiver or non-working back path, the broadcaster might waste a lot of valued bandwidth without a chance of success. A refinement of the pessimistic approach sends both negative acknowledgements (NAKs) and ACKs. Missing ACKs are still assumed to be NAKs.

5.4.2.6.2 Final Negative Acknowledgement Approach. In this approach, we assume that *no news is good news.* Whenever a receiver detects an incomplete package delivery, a NAK is automatically sent back to the broadcaster indicating a "request for a retransmission". No automatic ACK is generated when the file is successfully received. Missing NAKs are interpreted as successful transmissions.

The advantages are obvious, as the broadcaster does not waste bandwidth and stops retransmission when nobody responds with a NAK.

The disadvantage of the optimistic approach is that the broadcasters will not retry transmission if it does not receive any NAK. If a receiver is off-line for a short while and misses the first transmission, the broadcaster will not even know that someone has not received the broadcast.

5.4.2.6.3 Partial Acknowledgement Approach. The previous approaches treated the complete package as one unit. The whole package was sent again, when the user did not receive it intact in the first attempt. If reception was not confirmed for all receivers the NAK based approach can be refined to send a list of missing packets back to the broadcaster. The broadcaster collects this list and re-broadcasts only missing packets.

This approach reduces necessary bandwidth for the down-stream transmission considerably. Furthermore, it provides detailed information about what went wrong and when.

The approach requires more return path transmission than the other approaches, but reduces the downstream transmission requirements.

5.4.2.7 Last Resort Transmission in Package Delivery

Reliability in a package delivery system is, as described, achieved by a NAK collection and retransmission mechanism. A package delivery might consist of several retransmissions. After each retransmission, the broadcaster collects the NAKs from the package receiver within a specified time interval and, if needed, retransmits the missing portions of the package. The broadcaster sets the maximum number of retransmissions. However, if, after the last retransmission, the package is still not properly received, the package receiver alerts the end-user and asks permission to send a manual negative acknowledgement (MNAK). After sending an MNAK to the broadcaster a point-to-point connection is established to receive the missing packets.

5.4.2.8 *"NAK Implosion"* – *Managing Errors Statistically*

As discussed in the section on "reliable multicast" (the concept where multicasting was combined with acknowledgement messages and retransmissions) the broadcaster might be overwhelmed with incoming NAK messages. As most data broadcasting infrastructures are asymmetric in nature, the broadcaster cannot necessarily collect and collate NAKs along the downstream distribution path. However, there are three alternative ways of reducing this problem:

- *Timed NAK responses.* If each receiver sends the NAK messages immediately after end of transmission, the back channel (or the server handling them) might be overwhelmed with messages. A simple feed loop can reduce this probability. Each receiver waits a random interval until it sends the list of missing packets with the NAK message. This random interval should be shorter for receivers that have dedicated Internet connections available. The random timeout should be longest for receivers with slow, dial-up based back-channel. The broadcaster collects the NAK responses and broadcasts a message with the known list of missing packets as soon as it knows about missing packets. A receiver can go through this list and, if it has not sent the missing packet list back, can delete the packets that someone else has already reported as missing from the NAK list. By repeating this process the receiver hopefully empties or at least considerably reduces the NAK list by the time the random feedback-delay interval has expired.

- *NAK proxy servers.* Dedicated local NAK handlers might be set up to reduce the NAK implosion to the broadcaster. These local NAK handlers act as proxy servers for the broadcaster NAK handler. Receivers will connect to a NAK proxy instead of to the broadcaster. The NAK proxies then collect and collate the missing packet lists and send this list further up either to a next level NAK proxy server or to the broadcaster. NAK proxy servers might be installed as dedicated servers or at local Internet Service Providers with a fast back channel. If the data broadcasting distribution infrastructure is bi-directional,

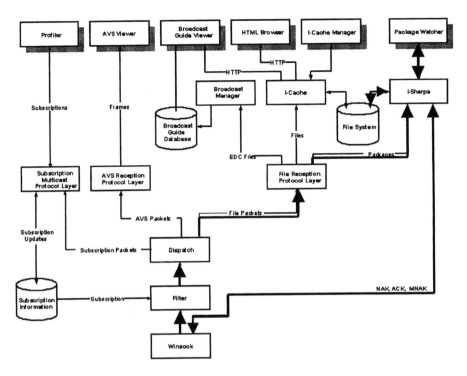

Figure 5.2 Example of a package delivery topology for data broadcast

like xDSL or cable, the NAK proxy servers should be combined with multicast routing facilities to reduce traffic even further. If a NAK proxy server is equipped with a broadcast receiver it can offer local handling of MNAKs.

- *Shut-up messages.* Implementation of shut-up messages might also reduce the problems associated with NAK implosion. A *shut-up* message is sent by a broadcaster if the broadcaster has determined that it will send the whole package again. Shut-up messages work with an optimistic, NAK based approach, where only a global NAK per file is collected, and not at a packet granularity.

NAK proxy servers and timed NAK response methods can be combined to reduce the risk of NAK implosion even further (Figure 5.2).

5.4.2.9 Package Viewers – How the End-User Accesses the Packages

The end-user would expect that a special user interface is available to handle package reception. These interfaces could inform the users of:

- Transmissions in progress

- Transmission statistics

- Transmissions needing attention

- Download through back-channel – MNAK handling

- Complete transmissions – package ready for viewing

- Activity to move packages from reception area

- Unwrapping of packet and communication with billing system

Some of this communication might happen in a secondary layer of the user interface so that you only see it if you are particularly interested.

The main package viewer should be launched through a broadcast guide entry. Since package reception is an automated process, user intervention should only be necessary when the automatic ACK and NAK handling does not succeed in getting the complete package. It should, in other words, only be necessary if the system has to switch to last resort in order to complete the transmission.

The package viewer is also the user interface for garbage collection. Since content delivered via package delivery is usually premium content, the system should not be allowed to just automatically delete packages based on age (as we know from many caching routines). The package garbage collection system should instead inform the user if packages are aged beyond the defined limit. The user should then decide if he wants to keep or delete the content.

5.4.2.10 Scheduling for Package Delivery

A content provider who wants to distribute packages via a data broad-casting network has to make a few key decisions:

- *Reliability requirements.* The content provider decides on the relia-bility requirements for a broadcast by selecting one of the previously mentioned protocols.

- *File and packet repetition parameters.* The content provider's choice of protocol can influence the price the broadcaster charges for a transmission as more infrastructure needs to be provided by the broadcaster.

Data broadcast management software might be able to help the broadcaster make that analysis.

5.4.2.11 Elapsed Time for Package Delivery

Package delivery raises one issue when planning how much time and bandwidth will be spent. Let's first go back to cached content and consider how that was broadcast. You assembled some files, entered them into data broadcast scheduling software, and this software would then calculate for you exactly how long it would take to broadcast this content assuming a given bandwidth. Now, for package delivery it is not quite as clear. Why not? Because in a package delivery model you spend an unpredictable time on:

- Waiting for ACKs, MAKs and MNAKs

- Collecting ACKs. NAKs and MNAKs

- Talking to receivers

- Communicating with ACK/NAK/MNAK proxy servers

- Sending missing packet lists

- Back-channel administration

This means that time and bandwidth required for retransmission is unpredictable. However, the broadcaster can use this idle time for broadcasts from the retransmission carousel.

5.4.2.12 Broadcaster Requirements for Package Delivery

The infrastructure to implement cached content delivery is not sufficient to provide package delivery. The major added requirement is the availability of a back-channel.

A logical infrastructure may let the content provider specify retransmission rate and bandwidth, while the broadcaster handles the back channel. After the last retransmission is done, the broadcaster tells all receivers that this was the last broadcast. End-users who still have not received the complete transmission now have to ask for the missing packets through the back-channel. The broadcaster will specify a dead-line for MNAK acceptance. Otherwise packages would need to be kept at the broadcaster forever. Infrastructure for the back-channel and support of the back-channel justify higher bandwidth prices for reliable delivery.

5.5 AUDIO/VIDEO STREAMING

You are running a data broadcasting service to set top boxes, car screens, handheld devices and PCs. It contains business news with text and pictures – updating around the clock. However, you have now decided to take the service to the next level. Whenever there is a really important news story breaking from the business arena you want to cover it with live video broadcast followed by sound-only discussions from a team of analysts, all in highest quality and all displayed within your users' brow-sers alongside the explanatory text. So how do you handle that? You need audio/video (A/V) streaming.

The concept of A/V streaming is hardly a mystery to anyone. We know it from radio and television. Some people will also know it from the Internet. A/V streaming in a data broadcasting environment is somewhere between what we know from those media. In terms of viewer experience it is very much like radio and television. The picture can be clear, sharp and big. It is updated at 30 frames/s, like television. Sound can be CD/DVD quality, or even surround sound. And it arrives live – there is no download time previous to viewing. However, from a technical aspect it is not terribly different from the way it is done on the Internet. The content is not analogue; it is divided into small digital packages which each have a header and a payload (content).

A/V streaming in a data broadcasting environment is:

- Best effort

- Time-critical

- Immediate view

A/V streaming implies that a pre-recorded file or a live event is broadcast. Unlike package delivery or cached content delivery, the file does not have to be completely received before it is played. However, contrary to cached content or package delivery, streamed content is not cached on the recei-

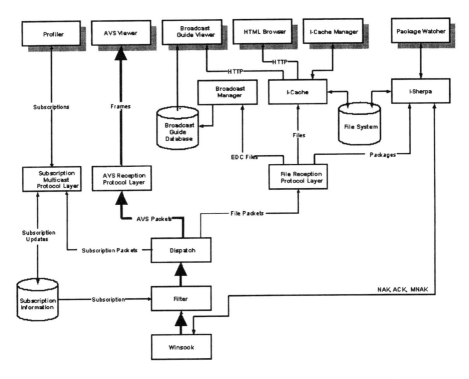

Figure 5.3 Example of A/V streaming software topology for data broadcasting

ver side, but is displayed as it is received (which has a key advantage, since it might otherwise take up huge amounts of hard drive space). But this means that the user has no control over when to start the streaming. He can only choose when to view the streaming file and this may or may not be in the middle of the broadcast – as we know from radio and television. This solution is near-on-demand scheduling (Figure 5.3).

5.5.1 Typical Applications for A/V Streaming

A/V streaming is extremely suitable for distribution of real-time information to a large audience. If A/V streaming is just considered as a stand alone service it does not offer any value over a business television or closed captioning system. However, A/V streaming combined with other delivery services provides real added value. Some examples of what it can be used for are:

- *Corporate announcements.* A/V streaming can be used as part of broadcasting corporate announcements. A video viewer shows the president of a company announcing the launch of a new product, while in a separate browser window a slide show highlighting the major points of the announcement is shown. Or a chief financial officer announces quarterly results of a company to analysts and institutional investors – perhaps via a network operated by a major investment bank for its key customers. While the A/V stream is used to distribute this particular information there can be simultaneous background transmission of earnings reports, financial statements and press releases to the viewers so they can browse at their leisure. This background transmission would then use cached content delivery as the format. Synchronisation of media streams and other content types can be achieved by using SMIL (Synchronised Multimedia Interaction Language), a W3 standard under development.

- *Daily stock market updates.* A/V streaming is also used for regular events. While the stock exchange is open, a financial institution broadcasts hourly financial status and trends, at the same time broadcasting press releases of companies mentioned during the broadcast as cached content. Analysis of these companies is delivered as a reliable broadcast and can be bought by the end-user.

5.5.2 Infrastructure Requirements for A/V streaming

The typical architecture for the reception of the A/V stream is one where A/V packets pass through the subscription filter and dispatch like any other delivery type. They are subsequently passed up to a viewer that is either stand alone or embedded in a web browser. Using this architecture allows the imposition of subscription control on a stream, while still providing all the advantages of HTML embedding of the viewer. Embedding the viewer inside a web browser allows for support of multimedia integration of different media types into one user experience. Architecture like the one presented allows integration with any streaming technology, as the stream player does not see a difference between a direct stream and a stream received through subscription filter and dispatcher.

In contrast to cached content delivery and package delivery, the stream is only available when it is broadcast. This makes it important that a user is alerted when a stream starts. This alert functionality is best performed by the broadcast guide, which may alert the user before the stream starts. If the user does not acknowledge that he is ready to receive the broadcast the filter will automatically discard the stream.

5.5.3 A/V Stream Event Types

Three different scheduling and playout methods allow covering of the A/V streaming requirements:

- *Streaming from file*. Audio and video are here encoded and stored in a file. The file is then scheduled for later broadcast and uploaded to the operator of the distribution network. At the scheduled broadcast time, the playout begins from the media file stored at the broadcaster's location. This scheduling and playout method is particularly useful when a media event has been pre-recorded some time before the broadcast is scheduled. The upload does not need to happen at the same time as the booking, but needs to happen before the playout. This independent playout does not require much bandwidth for the upload as long as the content is available at the broadcaster's location before the playout starts. For very large content the upload can even occur off-line by sending CDs or DVDs to the broadcaster via physical mail. However, streaming from file is a preferred method for pre-recorded content.

- *Streaming from file with propagation*. If content is pre-recorded, but there is no time to upload the content to the broadcaster, the ''streaming from file with propagation'' method can be used. In this method, the schedule is created and the bandwidth for the streaming broadcast is booked. Because of the booking, the broadcaster assigns the content provider a TCP/IP address and port to send the content to. At the given playout time the broadcaster accepts a connection on this TCP/IP port and replicates the stream for broadcast, encapsulating it with the correct subscription information. The advantage of this method is that the content does not need to be ready in advance of the scheduled playout time and that the broadcaster does not need to have sufficient storage for the complete A/V file. The disadvantage of this method is that we do not have any buffering at the broadcaster location, making it necessary to have guaranteed bandwidth available from the content provider to the broadcaster.

- *Live event streaming*. The live event streaming method is, as the name says, a vehicle for broadcasting streams covering live events. The broadcast is scheduled exactly as in the file propagation method. A video camera at the location of the event captures the event, and an encoder converts the video stream into an MPEG stream. At the time of the broadcast, this stream is accepted on a TCP/IP port at the broadcaster's location (assuming that the system is IP based). The stream is then wrapped into subscription packages and replicated onto the broadcast stream. The advantage of this method is that the content is not stored anywhere and the feed is directly broadcast. As in the file streaming with propagation method the lack of buffering at the broadcaster requires guaranteed bandwidth to be available from the content provider to the broadcaster.

5.5.4 Booking and scheduling considerations for A/V streaming

Scheduling streaming content for playout adds some unique considerations to the booking and layout system. The first one is a phenomenon called "audio and video thinning". The issue is this:

- An A/V stream is recorded with a certain frame rate and resolution resulting in unique bandwidth requirements.

- If the recorded bandwidth is larger than the scheduled playout bandwidth, or if the playout system needs to reduce bandwidth because of resource constraints or for economic reasons, we need to automatically change the bandwidth.

This is the process called "thinning". Video thinning can be achieved by reducing frame rate or changing resolution. Audio thinning can be achieved by reducing data rate or by switching from stereo to mono sound. When booking for an A/V stream event the content provider can set hints for the playout system to determine what thinning method to use if or when required.

Another issue is what is called "live feed pre-emption". When scheduling for cached content delivery the scheduling system needs any two of the three parameters:

- data size

- bandwidth

- broadcast elapsed time

to calculate the third. This is very simple. However, A/V streams are not scheduled based on size, but based on length of playout. If we do not book bandwidth for the appropriate time, the playout might be stopped when the scheduled time is over. Content providers need to make conscious decisions if it is acceptable that a stream is pre-empted or alternatively book bandwidth for extra time. A situation like this happens frequently when live events run longer than expected. A football game might go into overtime and viewers get naturally very upset when the broadcast stops before the final result is known (television people hate it when key matches move into overtime!). A booking and scheduling system might accept last minute booking for in-progress streams if bandwidth is still available.

5.6 DATA STREAMING

Imagine this: you are trading the stock markets. Not once or twice a year, but often, and for real money. You spend several hours a day tracing what goes on there. You want to know about how prices move, how much is traded, at which bid and ask prices, etc. You want to be able to track selected prices as charts, or to link their updates into your portfolio system (which you run in a spreadsheet). And you want all the information in absolutely real-time, so that you know the precise market prices when you execute a trading order. So you need a system with data streaming.

Data streaming or real-time data is a time-critical, best effort and viewed-when-broadcast type of delivery. Bandwidth for the streaming data is scheduled with the broadcaster, but the data are always broadcast out as soon as they are received from the content provider. So how does this differ from the basic concepts of A/V streaming?

A/V streaming is continuous streaming. Picture frames are sent at constant intervals. Data streaming, on the other hand, is discrete streaming. Data items do not arrive at constant time intervals, but need to be re-broadcast on a first-in-first-out principle. Data streaming applications cannot afford human intervention to re-format and re-distribute as this would destroy their real-time nature.

5.6.1 Applications for Data Streaming

Any application that requires point-to-multipoint distribution of discrete data in real-time is a candidate for data streaming. Some examples:

- *Stock tickers*. The paper ticker tape was the first data streaming application. Even on current financial channels on television, we see stock tickers as data streams.

- *Weather data*. Weather data are collected in many different places and broadcast to a large audience. Weather data might include temperature, humidity, visual conditions, precipitation, etc.

- *News wires*. The constant feed of headlines through a broadcast system can be thought of as a real-time data stream. Headlines might be stored in a carousel where the last 100 headlines are kept. From this carousel, the headlines are streamed out in a continuous fashion.

- *Sports data*. Live data about the speed of the ball, the pulse of the racing driver, the temperature, etc. which are distributed alongside a video stream to generate a multimedia sports channel would be a typical data streaming service. Another example is virtual racing where you race in a virtual NASCAR or Formula 1 car against real cars racing on the real tracks.

5.6.2 Granularity of Data Streams

Data streams are suitable for ''real-time'' coverage of events (delivered ''live'', when it happens). Data streams are also ''non-persistent'', i.e. each event gets overwritten by the next event.

Data stream services will often have high granularity and this raises issues. How do you define the entity of each sub-stream, or each channel within a data stream? Is, for instance, a New York stock exchange ticker one stream or is each financial instrument or stock symbol a unique stream? Is there one weather stream, one weather stream per country, one per town or one stream for temperature at one location? Let's take stock trading as an example. A stock exchange might distribute the following, constantly updated, information about a single stock:

- ''Long Name'' (the full name of the stock)

- ''Short Name'' (an abbreviated name of the stock)

- "Symbol" (the exchange's code symbol for the stock)

- "Trade" (last traded price)

- "High" (highest traded price during the day)

- "Low" (lowest traded price during the day)

- "Close" (last traded price during the day)

- "Volume" (number of shares traded in the last trade)

- "Total Volume" (total number of shares traded during the day)

- "Best Bid" (highest priced bid in an unfilled electronic order)

- "2nd Best Bid", "3rd Best Bid", "4th Best Bid", "5th Best Bid" (additional unfilled bids)

- "Best Ask" (lowest price asked in an unfilled electronic order)

- "2nd Best Ask", "3rd Best Ask", "4th Best Ask", "5th Best Ask" (additional unfilled asks)

This would be a total of 19 items updated during the day about a single stock. Now, since there are well over 200,000 traded financial instruments in the world, this means that a demanding user could have access to several million individual data channels.

This is actually the way it is approached if the system allows for the user to build "montages" on the screen that can update any individual item independently (which some systems actually offer). Such a high-granularity data streaming system might also allow the user to track any variable graphically, or link any variable into a spreadsheet where it may be subject to any real-time "rocket-science" analysis that this user finds useful.

However, you can easily imagine situations where you do not need to separate every single item, and this may speak for bundling together in larger groups.

5.6.2.1 Data Stream Bandwidth

Items sent in a data stream are in most cases numerous, but with each update being very small. Even a complex stock trade usually does not require more than a few hundred characters. A few examples are as follows. On one typical day (October 1, 1998) the New York Stock Exchange (NYSE) executed 720,615 trades. The average number of trades per day for January through September 1998 at the NYSE was 522,261. Constant streaming over a 7-h opening period results in a bandwidth requirement of less than 50 kbps. If we compare this to a good quality A/V stream requiring about 2 Mbps we see that bandwidth requirements for data streams are quite moderate – the real advantage of broadcast is not so much bandwidth as low transmission costs and guaranteed real-time access. A more detailed discussion on how scheduling, bandwidth and application requirements interact is given later.

5.6.2.2 Data Streaming Viewers

As each data stream has most likely a unique format it is necessary to allow for open viewer architecture. Viewers for data items can be stand alone applications, plug-ins in Internet browsers or existing applications. All viewers can use the existing receiver architecture to tie into a data stream.

Typical applications for presentation of streaming data are:

- *Chart.* A charting tool that visualises time series as they update and allows the user to manipulate and analyse the updating events graphically.

- *Map.* A geographical or political map that displays values of location-related time series through colour-coding. A roll-over feature could display the raw numerical data.

- *Quote Page.* Tables of live values updating figures, charts or text.

- *Newsreader.* A page where news headlines appear. The headline can be linked to the full story.

- *Ticker.* Scrolling prices or news headlines are presented and are linked to pages with more information.

- *Table.* Tables in a spreadsheet or another format showing the last value in the series.

Data stream viewers normally show only the last value in a given series. For some applications a historical view, collecting the value and time information, might be useful. If the content provider chooses to send multiple logical streams onto one data stream the viewer application has to support filtering on data items according to values. This is normally not a critical issue, as data streaming applications filter out items of interest. A stock trader might, for instance, have a screen that shows windows with only a few instruments.

5.6.2.3 When Streaming Data Coexists with Other Services

Data streams normally provide most value if they are linked to other services. Scrolling news headlines delivered as data streams can, for instance, be linked to a full news story delivered via cached content delivery or an example from financial information: while it is, important to see the current values in a financial application, understanding trend information is as important. Trend information needs to have all the historical data available. The solution is to capture the historical data from the data stream and store it locally, which means that selected streaming data is converted into cached data. The problem with this approach is that the data streaming approach is not reliable and there is therefore no guarantee that two receivers have exactly the same data.

The second problem is that an end-user might not be using the system all the time. Assume, for instance, that an individual investor goes on vacation. He would miss the historical data for this period, causing the user to use incorrect or incomplete data for his work. A more practical approach is here to store the historical data in a database at the content provider's location. This complete database is then delivered at regular intervals to all subscribers with reliable delivery.

5.6.2.4 Scheduling and Booking for Data Streaming

Several new issues need to be solved for data streaming scheduling and broadcasting.

The first is that to handle real-time data effectively a content provider cannot afford to have long latency between the capture of the data and the broadcast of the data. Real-time data has no standard format, and that means that a content provider aggregating real-time data from multiple sources has to convert these different data formats into a standard output format for viewing. These aggregated streams should automatically be sent to the broadcaster by a "feed-handler".

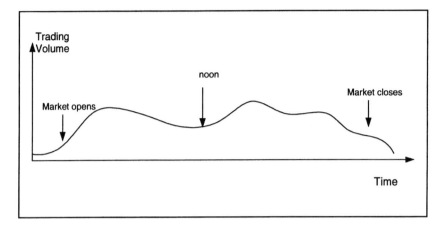

Figure 5.4   Potential content flow from a stock exchange

Two different architectures, similar to the ones illustrated in the A/V streaming section, can be used for data streaming feed-handlers:

- *Live data streaming from the broadcaster.* The feed-handler is located at the broadcaster site to decrease latency in broadcasting.

- *Live data stream propagation.* The feed handler translates the raw data into a data stream at the content provider location. The stream is sent to the broadcaster via a network. A network with guaranteed bandwidth is required to support this architecture.

Another issue to consider is the scheduling implications of non-constant data rate of data streams. The stock market example from before serves to illustrate this issue. Since trades do not happen at constant intervals, but are rather concentrated around opening and closing hours and special events, it is not practical to just reserve constant bandwidth for such a service. The trade volume over a day might look rather like Figure 5.4.

If the content provider just reserves the average bandwidth, two things can happen. Either data items are lost or the content provider needs to buffer items until bandwidth is available again. For delivery of time critical data, neither solution is acceptable. In this case the content provider will need to allocate extra bandwidth, so that buffering is reduced to an acceptable level. This leaves extra bandwidth available, though. Data streaming applications use in many cases additional services to deliver complete updates of historical data.

A combination of multiple delivery types scheduled into the same reserved band creates an acceptable solution. The content provider schedules the data stream with a given, fixed, bandwidth, and schedules additional data for either cached content or package delivery into the same bandwidth. The scheduling system accepts this back-up data with lower priority then the data stream. The broadcaster has to handle the different delivery types and use the priority to manage the playout bandwidth accordingly.

5.6.2.5 Managing Broadcast of Streaming Data

Data streams are, as A/V streams, broadcast as long as the schedule instructs to broadcast. The maximum data amount broadcast depends on the allocated bandwidth and the duration of the broadcast. The actual data transferred may be less, if the input data and the lower priority fill data does not require the full bandwidth.

The broadcaster for data streams assigns the data stream the highest priority. If bandwidth is available, the broadcaster also checks if any lower priority, non-streaming content is scheduled into the same band. Using free bandwidth to mix in lower priority services requires the broadcaster to react very rapidly to changes in the input data stream to change the rate on the lower priority stream. If the stream data capacity requires more bandwidth than allocated, the broadcaster must either buffer the data stream items, or, if there is general bandwidth available, use the available bandwidth to overcome short bursts of data.

5.7 BROADCAST GUIDES

Television programme guides, either on-line or printed on paper, are used by TV viewers to find out what programmes are broadcast on a given station at a given time. Similar mechanisms have to exist for data broadcasting. These can be called ''broadcast guides''.

A broadcast guide is a tool that lets the broadcaster/content provider inform all users of upcoming events and broadcasts. Guides can be generated that match a programming schedule. The broadcast guide should contain all broadcast requests, the associated content, and information submitted by content providers through the booking system. Specialised content broadcasters (playout processes) should be responsible for transmitting content and programme information as specified in the corresponding schedules.

Broadcast guide data should be delivered as a best-effort and time-critical distribution service. Best-effort delivery is adequate, as it is not critical if an item describing a programme is missing. The broadcast guide items are time-critical, however, as they have to be available before the broadcast described by an item starts.

5.7.1 The Media Context of the Broadcast Guide

Data broadcasting services are probably just one of the services offered to an end-user on the same receiver (PC, set top box, mobile device, etc.). In this case the broadcast guide itself can be accessed from some other, top level guide. This can be a login screen on a PC, a desktop on a PC, or a home-page on a set top box.

A data broadcasting system requires more than just an interactive-TV-like electronic programme guide. With the broadband data broadcast infrastructure it is also possible to make broadcast data more informative. The main goal for the broadcast guide is to support user-friendly navigation to content. Finding a channel/service as well as individual broadcasts is the goal of a broadcast guide. Top level groupings for services, is just one way of achieving this goal. With the number of services available, it is not realistic for a user to browse through all the different broadcast channels. The guide data needs to be narrowed down to user preferences first to find services and entries of particular interest. Finding programmes and navigation to broadcast data has to be content driven.

5.7.1.1 Locating Content from a Broadcast Guide

The end-user's goal of using a broadcast guide is to find relevant content. Relevant content is personal content. Personalisation of the broadcast guide is either initiated by the end-user or initiated by the system. Broadcast guide entries should provide at least the following properties for a broadcast:

- *Availability*. When is the data item available for viewing and for how long is it available?

- *Type of delivery*. What distribution type is used for a broadcast item?

This influences user behaviour. For cached content the end-user knows that the content will be available for browsing after the indicated time, for streaming the user needs to be present when the broadcast starts, for package delivery a user has a certain time after the broadcast to collect missing data manually.

- *Ratings*. Broadcast data items should be rated with standard rating criteria like age group, adult content, violence, etc.

- *Language*. The natural languages in which the broadcast is available.

- *Title*. Each data broadcast item has a title.

- *Description*. Every broadcast item has a brief description.

- *Interest group*. Top-level interest groups like news, sports, movies.

- *Size*. Size of broadcast.

- *Preview*. A movie clip, still pictures or text previewing the event. Preview items can contain multimedia content.

- *Origin*. Geographical, political origin as well as possible URL pointing to additional content.

- *Programme information*. Information like a serial number, sequence number, volume, publication date.

- *Content*. Provider-specific information.

The values for these properties are defined by the content provider and searchable by the end-user. An end-user can actively search through a broadcast guide using any of the properties listed. The system should allow users to combine search criteria into logical queries, and it should support explicit and implicit queries:

- Explicit queries are executed whenever the user starts the query.

- Implicit queries are executed on behalf of the user. Whenever new broadcast guide entries are added the system will execute the queries and update the results for viewing. Implicit queries are just one way of providing intelligent personalisation.

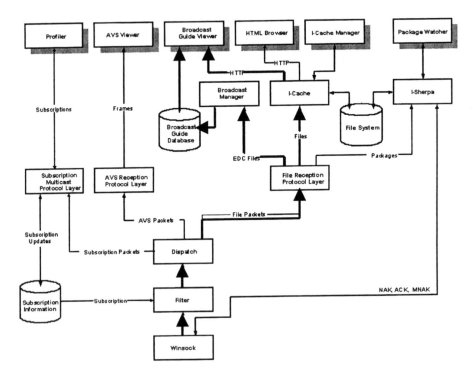

Figure 5.5   Example of broadcast guide topology for data broadcasting

5.7.1.2 System Driven Personalisation and the Broadcast Guide

Broadcast guides should share technology with the intelligent caching described earlier. The system can use this technology to *look* at the broadcast guide items and alerts the user of broadcasts of interest. Using intelligent agents as well as end-user personalisation, the system has enough information to create a personal daily front page when the user accesses the broadband system. Using templates from the content provider, the results of broadcast guide personalisation and the local content assembly techniques described in the cached content delivery section, the system can build personal front pages. After the user has found the content he is looking for, the user can navigate to the item of interest (Figure 5.5).

5.7.1.3 Navigation Issues of the Broadcast Guide

Navigation in a broadcast guide means:

• Browse HTML based content for cached content delivery

- Open the package watcher for reliable delivery

- Start the streaming viewer for A/V streaming

- Open the data streaming application

However, the mere flexibility of data broadcast technology can raise an issue. Assume that there are hundreds of channels, and that each user subscribed to only some of those. How do you ensure that they are not constantly diverted to content that they cannot access because they are not entitled to it? As described before, two factors determine if a user receives a service:

- *Is the user subscribed for the channel?* Did the subscription system send out an entitlement message?

- *Is the user enabled for the channel?* Did the end-user decide that he/ she wants to receive the content or not?

This is handled through the way in which the broadcast guide is communicated to each user. Firstly, it can be distributed in different ways:

- On a shared basis

- As a free-to-air channel

- On a service controlled by subscription

In the first case, we use the broadcast guide to advertise a service or an individual programme. The content itself is either available to all end-users as part of the basic system services, or the user can subscribe to this service. Broadcast guide data on free-to-air channels are used when every user can subscribe to this service, even if it is for a separate fee. It does not make sense to send broadcast guide data on channels available to everybody if the content is not available to everybody. A company might only make its channels available to employees, customers and suppliers. The company does not want to allow anybody outside this community to even see what content will be broadcast.

Broadcast guide entries share implicitly the rating with the content they describe. An item describing adult content might contain a preview and description that might not be suitable for everybody. The content provider can explicitly overwrite the implicit rating. Several key technology decisions allow these goals to be implemented.

5.7.1.4 Broadcast Guide Technologies

The goals outlined above require flexible, extensible and active technology. Advance and last minute notification of upcoming events is one of the requirements for the broadcast guide. This means the broadcast guide needs to be always up-do-date. A paradigm for broadcast guide reception is required.

Broadcast guide data for many channels/services can be large. As described earlier, explicit and implicit filtering of guide entries is a requirement. A local database allowing for explicit and also agent based searches is best suited to fulfil these goals. This leaves us with the extensibility goal. Broadcast guide entries contain many predefined properties as shown above. These entries have different types, text, number, lists, picture, sounds, and video-clips. HTML is therefore a natural candidate for specification of broadcast guide entries. HTML entries describing broadcast events also make it possible for content providers to define a private, unique look and feel for a guide. Innovation is therefore possible. One of the goals of the broadcast guide is extensibility. HTML itself does not offer this capability. HTML has a fixed, limited set of tags to describe content and formatting of content.

HTML is an application of SGML, the Standard General Mark-up Language. SGML provides all the extensibility required for a broadcast guide, but was thought to be too complicated as a simple Internet mark-up language. XML, the eXtensible Mark-up Language, is a subset of SGML, a sibling to HTML, which is simple enough for web delivery, but made powerful enough by providing user definable tags.

Up to now we described the goals, reception and viewers for a broadcast guide. However, we should also consider who decides what to send out and when to broadcast guide entries.

5.7.1.5 Roles in Creating the Broadcast Guide

As described earlier, a data broadcasting lifecycle has at least four distinct roles:

- Content provider
- Community administrator

- Broadcaster

- End-user

Let's take a brief look at the roles of each of these players:

- *Content provider.* The content provider is responsible for creating the broadcast guide entries for individual content and can also add information about itself. Describing the business and contact information are just two possible entries the content provider can use. The content provider works under guidelines developed by the community administrator.

- *Community administrator.* The community administrator is responsible for describing a channel/service and defining policies for the content providers. Policies include bandwidth (how much to use for a guide entry); scheduling policies (how long in advance should a broadcast guide entry be sent); guide content (standard conventions the content provider will need to observe for layout, usage of multimedia, usage of ratings, custom defined tags); and free to air or subscribed (the community administrator also makes the choice between providing the broadcast guide on a free to air channel, so everybody can see a description of the channel and the broadcasts; the alternative is to send the broadcast guide data on a dedicated service within a channel owned by the community). The content provider and community administrator schedule the broadcast of the broadcast guide data with the broadcaster.

- *Broadcaster.* The broadcaster becomes the aggregator of broadcast guide information for information broadcast on free to air channels. The broadcaster will allocate broadcast guide data bandwidth for the free to air broadcast guide. The aggregated broadcast guide information is broadcast as best effort delivery. The broadcast guide data is sent out repeatedly from a guide carousel. Even low broadcast-guide bandwidth delivers a large amount of data. A constant 5 kbps stream, for example, delivers about 50 Mbytes of data per day. Assuming the average size of a broadcast guide entry of 1 kbyte and with five repetitions for each entry, the system could send out over 10,000 entries just on the free to air broadcast guide. The broadcaster sets policies for the community administrator to define how much bandwidth can be used for broadcast guide information on the shared broadcast guide bandwidth. The broadcaster does not control bandwidth for the subscription based broadcast guide, as this bandwidth is scheduled and paid for by the community.

- *End-user*. End-users can use the guide to search for different programming items or change the settings of the guide so that the items they generally are most interested in appear on top of the lists.

5.8 LINKING BROADCASTING SERVICES TO THE INTERNET

We have now looked at five different basic broadcasting formats:

- Cached content

- Package delivery

- A/V streaming

- Streaming data

- Broadcast guide data

All of these enable a content distributor to broadcast digital content to the end-users. One good question is how this can be combined with the Internet. How do the two media merge?

5.8.1 The Three-Layered Hybrid Medium

One of the strongest applications of data broadcasting is to use it as building blocks for a three-layered hybrid (or hyper) medium, where the three layers are "pure broadcast", "walled garden" and "the Internet" (Figure 5.6):

1 *The broadcast layer.* What you should experience at the highest level in this combined medium is something that is very close to television (which was, after all, a great commercial hit: there are at least 1.4 billion units installed). This means a layer which is most of the time dominated by streaming video, streaming data and CD quality sound. However, there should be a direct connection from this layer into the two other layers.

2 *Walled garden layer.* The walled garden layer should contain interactive content that has been broadcast to the hard-drive. The user should be able to explore this content, which may be very media-

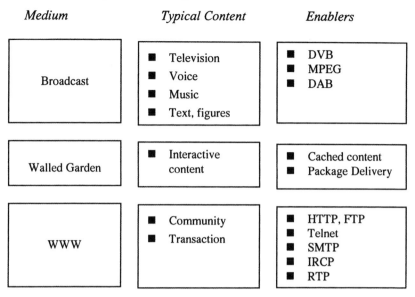

Figure 5.6 Three layers of a converged medium

rich and which will be instantly accessible. There should be direct connections from areas in this walled garden into the Internet.

3 *Internet layer.* Access from the broadcast layer and the walled garden layers into the Internet should happen in two different ways: initiated by the user or initiated by the broadcast content (more about that later).

The basic navigation method in such a system will be based on the user clicking on buttons or links, entering URLs or launching web searches. Broadcast content, walled garden content and Internet content should often be available alongside each other and simultaneously within a single interface. Users should thus not bother about the technologies behind it; what matters is that you get a single, seamless medium that combines the speed, drama and emotion of broadcast with communities, diversity and transactions of the Internet. However, there is an additional navigation method possible. We can call this "Bouncing".

5.8.2 Creating "Bouncers"

Imagine that you are listening to a digital music flow. It is just running in the background while you are doing something else. Suddenly you hear a song that you really like. So you hit a button and a purchase form for the

Table 5.3   Examples of ''bouncers''

Broadcast content that triggers an automated search (video)	Auto-triggered Internet dial-up
General news stories (text, stills)	*Auto-WWW search for related information.* The search result is subsequently broadcast and appears as an attachment to the broadcast news story *Auto-search into digital encyclopaedia.* Phenomena in the story that appear in the encyclopaedia can be highlighted by a click and the relevant encyclopaedia content accessed directly from the hard-drive *Auto-search into book-site (such as Amazon or B&N).* The search lists a number of books about subjects in the news. Information about these books is automatically broadcast and can thus be accessed directly from the hard-drive
Sport news (news, stills, video)	Same as above, but also auto-links into sports database
Financial news (news, quotes, graphs, video)	Same as for news, but including focused links into relevant databases such as Edgar and relevant corporate websites
Background music (CD quality)	Automated link into music shopping site. A digital form to purchase the album of any track playing is automatically filled out and broadcast to each user. It is thus ready for adding to the electronic shopping basket

relevant album pops up. The form is already filled in: it has the name, description and price of the album as well as your own details. Click ''accept'' and it is added to your digital shopping cart. Once you have added 10–15 albums (perhaps over a couple of weeks) you make a review, cut a few out, and then you send the purchase form via the Internet. Four days later you get a package with the albums you bought.

So what happened here technically? The broadcaster is background-downloading information about the album playing. A server at the broadcast head-end extracts this information and sends a "bouncer" into a music e-commerce site on the Internet. What bounced from the Internet was a digital purchase form for the album playing right now. This form was then automatically pushed out to the hard-drives of all end-users, where it was stored in the walled garden. Bouncers can be used in a number of situations, as show in Table 5.3.

The combination of broadcast for instance consumption, broadcast for delayed consumption, the Internet, and smart bridges ("Bouncers") between the three creates the ultimate medium.

Facing the Music

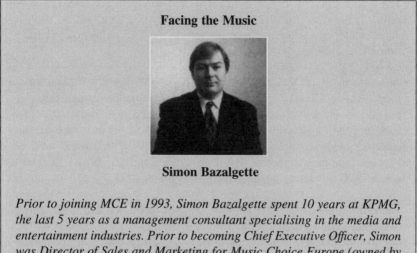

Simon Bazalgette

Prior to joining MCE in 1993, Simon Bazalgette spent 10 years at KPMG, the last 5 years as a management consultant specialising in the media and entertainment industries. Prior to becoming Chief Executive Officer, Simon was Director of Sales and Marketing for Music Choice Europe (owned by BskyB, Time Warner and Sony). He was responsible for building cable and satellite distribution for the Music Choice™ service and for the marketing and brand positioning of the service to consumers. He was also responsible for business affairs, including copyright and regulatory licensing and contract negotiation, and advising on corporate strategy.

Data broadcasting – this is undoubtedly a subject on which there is much more heat than light, particularly in the music industry. At the mention of an "on-line music service" many in the music industry immediately start explaining how such services will be the death of the record company or perhaps how it is the domain of copyright pirates.

Convergence is another term that seems to inspire sharply divergent opinions. To some, convergence will certainly take place and everything will become controlled by the computer industry or the telephone industry. To others, convergence is a myth because the PC and the TV will remain separate, one used for "lean-forward" use and one for "lean-back".

Into this minefield let me try to bring Music Choice's™ view of these issues. Music Choice Europe programmes and broadcasts 50 channels of music across Europe. We have a sister company in the US that does the same across North and South America. Each channel represents a difference genre of music, from Classical Baroque to Classic Rock, without interruptions from DJs or advertising. The broadcast includes a data stream which provides the title of each track that plays, the artist, the album from which it is taken and the record label.

The Music Choic™ service uses digital broadcasting technology, and is primarily broadcast via digital platforms, received by the consumer using a digital set top box which is connected to both the TV and the hi-fi to ensure near-CD quality sound.

Although the main focus is on cable and satellite digital TV based platforms, Music Choice™ is open to use any distribution system that can provide an attractive service to consumers at the right price. This can include FM transmission to provide a limited ten-channel service on cable, or DAB technology to provide a terrestrial digital radio service. However, the most attractive new distribution is likely to be to consumers via their PC, obviating the need for the set top box.

There are strong correlations between the use of a PC at home and listening to music. A multi-channel music service to the PC is therefore likely to find a ready audience and to be an important part of data broadcasting offerings. This medium also offers greater possibilities than cable and satellite today for added data and interactive services such as CD sales. There are other differences that need to be addressed. The PC and a TV (or a TV set top box) tend to be placed in parts of the house with quite different uses. Music is likely to be used differently in a study than in the living-room, and the programming may need to be adapted to take account of this.

In addition, the music industry is cautious about providing music to the PC, and needs to be convinced that an audio-streaming service is merely an extension of broadcast, albeit with some additional protections required to prevent piracy.

Therefore, providing a music service such as Music Choice™ via data broadcasting platforms is still at the developmental stage; however, in the medium to long term it has the potential to become as important as traditional cable and satellite media. It is to be ignored at our peril!

Simon Bazalgette

6. Implementing a Data Broadcasting Platform

"It has always seemed to me that the most difficult part of building a bridge would be the start"

Robert Benchley

We have now looked at the five basic categories of data that are relevant for data broadcasting. The next step to consider now is how to create a complete communication platform for data broadcasting. We can start by considering the basic question: what is a communication platform really? What does it need to be able to handle, other than basically moving the five categories of content we just described from A to Z?

One answer to this is that it needs to be able to support all aspects of data broadcasting when you run it as a professional medium, and all aspects of it when you run it as a business. Let's take a closer look at what this really means.

6.1 FACTORS DETERMINING BASIC PLATFORM REQUIRE- MENTS

A good starting point for this consideration is to consider the data broadcasting ecosystem again (Figure 6.1).

A data broadcast platform needs to support the ways each player wants to use data broadcast as a medium and as a business. The ecosystem participants will each play their specific roles in the process flow, and it is the fulfilment of these roles that defines the technical requirements for a data broadcasting platform.

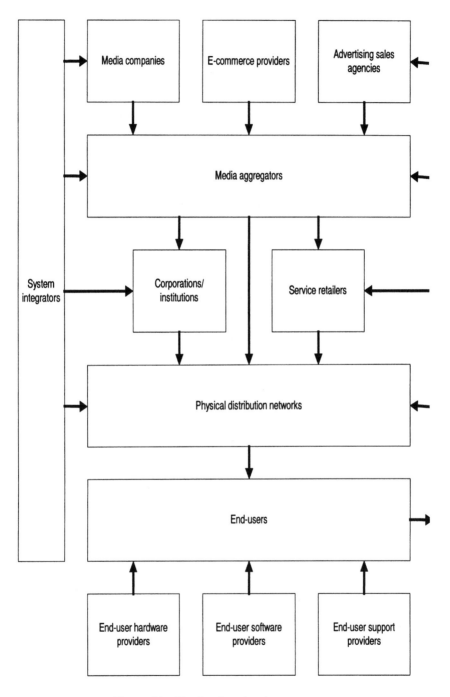

Figure 6.1 The data broadcasting ecosystem

6.1.1 The Role of Content Management

Media companies, organisations/institutions and *media aggregators* may have a requirement for managing content specifically for data broadcasting purposes, since they need to:

- Automatically re-format and compile content for data broadcast playout

- Track changes in the content flow

6.1.2 The Role of Community Administration

Organisations, institutions, media aggregators, e-commerce providers and service retailers may play the role of community administration. This means that they will:

- Manage one or more communities

- Be responsible for user registrations and subscriptions

- Maintain the services/applications assigned to the community

- Assign services to the content providers

- License one or more communities to subscribe to their services

6.1.3. The Role of Network Administration

Companies that operate *physical broadband networks* (such as satellite, cable, terrestrial or xDSL) have the following main tasks:

- Provide multimedia distribution services to content providers through the Booking System

- Execute broadcast requests (schedules) as specified by content providers

- Provide community administration services to a community administrator

- Provide access to reporting and billing information to content providers

- Administer the broadcast system though a monitoring application and a database configuration tool.

6.1.4 The Role of Advertising Management

Finally, there is a role of inserting and measuring the effects of *advertising*. This includes action to:

- Insert advertisements in the content flow

- Measure end-users' access to each advertisement

- Provide reports to advertisers

6.2 LISTING THE BASIC FUNCTIONALITY REQUIRED

The following is a list of the basic functionalities that are required in order to fulfil the four basic roles of the value chain players (managing content, communities, bandwidth and advertising):

- *Subscription management.* Assuming that you transmit content that is not meant for everyone, then you need to have a subscription management system in place. Assume, for instance, that you want to transmit live data from New York Stock Exchange to selected subscribers. This is expensive information, and you obviously want to be able to handle subscriptions through the system.

- *Data encryption and security.* You need to ensure that data meant for one group cannot be viewed by another. We could imagine two car manufacturers both running a distance learning system over the same satellite. It would, in such a case, be a disaster if employees in one company could participate in the sales training provided by the competitor.

- *Bandwidth management.* Content in a broadcast system is not sitting like website content on a server and waiting for individual users to access it one by one. It is being transmitted at a given time to all its users, and the transmission will require different bandwidths during the day, depending on what goes out when. This means that there is a need for planning how the available bandwidth is being used.

- *Scheduling and booking.* The transmission issue also raises questions about transmission timing and access to bandwidth. How can you be sure that the bandwidth is available on the network when you want it? Perhaps someone else has booked it all at the time when you need it. You need to be able to book the bandwidth you need in advance, so that you can be sure that the transmission will actually go through.

- *Billing.* Anyone who wants to distribute premium content will of course require the presence of a billing system.

- *Reporting.* Any content provider, whether it is for a corporate in-house system or for commercial distribution of media, will need to be able to track who it was sent to, so that they can pay for the copyright.

- *Media object tracking.* Whenever you manage media you need to be able to track how it has been flowing: what came in, what came out, when it was changed, etc.

- *Intelligent content compiling.* The concept of broadcasting will in most cases imply a continuous flow of content rather than a static database, as we know from most websites. This means a requirement for tools that can automatically process the data for this continuous flow, just as there are standard tools to automatically build websites.

A data broadcasting platform must have modules that can support each of the tasks listed above. Other than the specific functionality mentioned, there are some general features that are desirable in such a data broadcasting platform:

- Hardware independence (run on PCs, set top boxes, car communication systems, hand-held devices, etc.)

- Network independence (run over satellite, cable, DSL, digital terrestrial, mobile, etc.)

- Browser independence

We shall address these issues after we have looked at each of the commercial roles and the technical requirements they generate.

Table 6.1 provides an overview of the different players in the data broadcasting value chain (other than enablers) and their basic requirements for each of the functions mentioned above.

Table 6.1 Data broadcasting platform features required by different value-chain players

	Subscription management	Data encryption and security	Bandwidth management and scheduling	Bandwidth booking	Billing	Reporting	Media object tracking	Intelligent content compiling
Senders								
Media companies	✓	✓	✓	✓	✓	✓	✓	✓
Corporations / institutions	✓	✓	✓	✓		✓	✓	✓
Advertising sales agencies						✓	✓	✓
E-commerce providers		✓			✓			✓
Distributors								
Media aggregators	✓	✓	✓	✓	✓	✓	✓	
Physical broadband networks			✓	✓				
Service retailers	✓	✓	✓	✓	✓	✓		
Support providers								
Receivers								
End-users	✓	✓						

6.3 SUBSCRIPTION MANAGEMENT

It is important to have a Subscription Management System (SMS) that lets you manage and maintain subscription data, manage communities tied together by common services, and in some cases, even license services to outside communities. An SMS will:

- Establish customer accounts

- Assign subscription rights to those customers (end-users)

- Provide a relationship between a channel (or a service) and an end-user's receiving hardware/software

A data broadcast platform needs to support these functions.

6.4 DATA ENCRYPTION AND SECURITY

We have previously compared data communication to the postal service. The postal service today is perceived as a "secure" delivery mechanism. Mail sent in a sealed non-transparent envelope is assumed to be read only by the designated receiver. However, sending non-encrypted data over the Internet or over a broadcast platform today is more similar to sending a postcard: someone other than the recipient might read it.

However, data encryption – the use of cryptography – makes it very difficult (if not impossible) to read the content of a data transmission for somebody other than the designated receiver. Encryption translates data into a secret code and is the most effective way to achieve data security. To read an encrypted file, you must have access to a secret key or password that enables you to decrypt it.

Unencrypted data is called "plain text"; encrypted data is referred to as "cipher text". There are two main types of encryption: "asymmetric encryption" (also called public key encryption) and "symmetric encryption" (simple key based cryptography).

6.4.1 Encryption and Conditional Access Systems

When cryptography is applied to communication to increase security, every message is *encrypted* before transmission. The message is thereby transformed in a way that can be understood only by the designated receiver. This receiver can *decrypt* the message using a cryptographic key transforming the message to a readable format (this will typically happen automatically).

A simple form of user authentication is a user name and password. Public key cryptographic systems, described below, provide a more sophisticated form of authentication that uses an unforgeable electronic signature. Two mechanisms are used today in communication systems: *simple key based cryptography* and *public key based cryptography*.

- *Cryptography based on simple keys.* This mechanism uses the same key to encrypt and to decrypt messages. For this reason it is also called *symmetric key cryptography.* The key has to be exchanged between receiver and sender on a secure channel before the actual communication can start. Simple key encryption systems are known to be "not very secure", but are simple and fast.

- *Cryptography based on public keys.* This mechanism uses two distinct keys to encrypt and decrypt messages. A so-called *public key* can be distributed to everybody and is used to encrypt messages. A second key, called a *private key*, has to be stored in a safe place and is used to decrypt messages. As illustrated in Figure 6.2, a pair of public and private keys is required in order to transmit information from sender to receiver.

Data broadcasting content can be encrypted to any other digital content. In a broadcasting environment a secure transmission has to be guaranteed between one sender and multiple receivers. Because a message is simultaneously sent to everybody, it has to be encrypted and decrypted using one single key (or pair of keys). A public key system relies on the assumption that a receiver creates a pair of keys and publishes the encryption key.

6.5 SCHEDULING AND BOOKING

A third important area that a data broadcasting platform needs to be able to address is *bandwidth management* from the sender's point of view. Bandwidth is the maximum amount of data that a connection can transmit in a given period of time. The amount of information that a communications channel can carry is directly proportional to a channel's bandwidth (bandwidth is measured in bits per second, "bps").

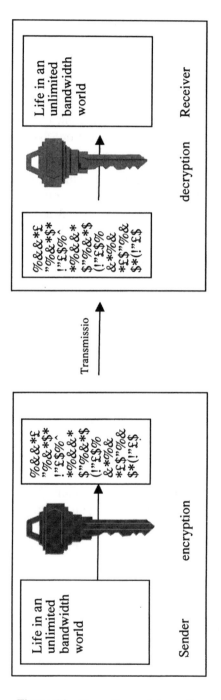

Figure 6.2   Encryption and decryption

Let's first consider a well-established business that has a similar problem: television broadcasting. A television broadcaster will have a team of people that creates a schedule for the broadcast. This schedule will be planned some time in advance and will ensure that the screen does not go black at given times, that advertisements can be inserted at the appropriate places, etc. Some of the actual transmission according to this schedule will be handled by a mechanical machine that will take video cassettes (or DVDs) from shelves, put them into VCRs (or DVD players), and play them out at the scheduled times. This means that the actual transmission, at least some of the time, is happening automatically following a schedule that was made previously.

One thing that makes the planning of traditional television bandwidth reasonably simple is that the bandwidth usage follows a linear scheme: it is video all the time, and the reception is simultaneous with the transmission.

The bandwidth management issue for data broadcasting is more complex, as it involves broadcasting combinations of five classes or kinds of media:

- Cached content delivery

- Audio/video streaming

- Package delivery

- Real-time data streaming

- Broadcast guide data

The derived complications are:

- Each media category may have different sizes.

- You may want to transmit several parts simultaneously.

- You may transmit for synchronous viewing or for store-and-forward.

- You may require guaranteed real-time transmission or accept some slack.

- You may desire that parts of the content are retransmitted X number of times while other parts should be transmitted only once.

- You may desire that some is delivered as guaranteed packages while the rest is best effort.

In order to plan bandwidth with such a complex set of options you need a graphical planning tool that makes it possible to get an overview of the options and visually see how you may "pack the content into the bandwidth pipe". One way to do this is to represent each part of the content as a block in a diagram, where, for instance, the height of the block represents bandwidth, the length represents time, and the colour represents transmission class. This makes it possible to play with options in a scheduling tool before finally choosing an optional solution. The ideal tool will then help you fill in spare bandwidth with desired retransmissions, so that the overall usage is optimised.

6.5.1 Features of Scheduling Tools

A schedule is a plan that shows what content will be transmitted during which time intervals, to which user groups, using which networks and which quality of service. The schedule itself is initially only isolated, created and owned by the sender. What this sender ideally wants to be able to do when building this schedule is:

- *Create a schedule.* With this request, the sender asks to check the necessary bandwidth availability to broadcast content

- *Delete a schedule.* The sender asks for the deletion of one of his schedules

- *Modify a schedule.* The sender manager asks to modify one of his schedules

- *Define a standard cut-off rate.* Define a time-limit before playout that defines the latest time where changes in the schedule can be accepted

- *Reload all schedules.* This request updates of the local schedule database with all the schedules booked on the server database

However, managing the bandwidth is not only a task for the sender. The sender is going to try to transmit through a pipe which is presumably shared by many users, and this means that you need to ensure that your part of the bandwidth is actually available – that your plan can be fulfilled. This is a "booking" issue.

6.5.2 General Criteria of Bandwidth Booking Systems

When you pick up your normal phone or cell-phone to make a call, then you will expect a so-called ''dial-tone''. The purpose of this tone is to inform you that bandwidth has now been reserved for you from where you are to the closest central switchboard, and you will not start entering the number or start attempting any transmission of anything until you have heard this tone. This is a very simple bandwidth booking system.

The traditional phone bandwidth booking systems have other basic features. You may book bandwidth in advance on a network through a so-called ''leased line'' (also called private line, or full period line). This booking means that you know in advance that the bandwidth is there when you want it. When you use a cell-phone, then you will often be able to see on the display how good your connection is. This is additional booking information. And there are phone systems where you can see on the display how much you are paying when you are connected. The price may be different at nights and weekends, and it will surely depend on the distance of your call, and this will all be communicated to you.

A more complex situation is where you use a modem and this modem sends information looking for another modem at the other end. The information that goes out here is:

> *''I am this-and-that kind of modem and I am looking to send at this-and-that speed! Who are you?''*

The other modem will understand what this modem said and will communicate back:

> *''I am so-and-so modem, and I can confirm that you can send at this-and-that speed''*

This is a so-called ''negotiation'', or ''handshake''. A good scheduling/ booking system is also required in a data broadcasting system, and this should be able to handle this ''negotiation'' between sender and network. This should ideally provide the following features:

- *Convey the desired bandwidth allocation schedule from sender to network.*It should send information about how the sender intends to use bandwidth to the network as a request.

- *Communicate system status in advance.* The sender should receive clear information back from the network about the availability of the desired bandwidth.

- *Support different bandwidth classes.* The system should support a range of alternative bandwidth classes and prices.

- *Support cancellations.* It should be possible for a sender to cancel a booked bandwidth and for the broadcasting network to use that information for scheduling the use of its bandwidth for overall optimisation.

- *Support last-minute booking.* Some content providers may want to book bandwidth at the very last minute, and this should ideally be supported as well.

- *Communicate alternative bandwidth options.*The network should ideally be able to send additional information about alternative bandwidth options that may be relevant for the sender.

- *Optimise network bandwidth allocation.*The network should have tools that optimise its use as it gets information about different schedules.

In any business, the booking of a service should ensure the customer that the service will be provided with all the features he has requested. Many additional features can be offered in a booking process, as long as they do not reduce the benefits for both consumer and business as arranged by the booking agreement.

6.5.3 Managing Booking Through a Transmission Chain

Everyone knows the situation where you do manage to get a dial-tone on your phone, but you cannot reach the receiver anyway. The reason for the failure may be:

- His phone is ringing, but he does not pick up

- His line is busy

- The links to his country or area are over-stretched, so you cannot get through

- The number you dialled does not exist or belongs to a different person

What you encounter here are complications relating to the chain of communication. The first part of the chain (your dial-tone) might work, but something goes wrong further down the chain. This may not be because there is something wrong with the basic architecture with the system – the fact that you get information about what is wrong actually tells you that the system is designed to deal with the issues, even if it could not solve the actual problem.

A data broadcasting scheduling and booking system has to be able to handle similar issues, even if the final result in some cases may be the same – that the sender cannot get the information through for commercially acceptable reasons. The issues in a data broadcasting system are:

- *Roaming.* Traditional switching networks are interlinked through roaming agreements. These agreements enable a user to communicate through a chain of telecommunication networks. A good data broadcasting platform is capable of supporting such roaming agreements so that data can "hop" seamlessly from network to network and so that a bandwidth request and bandwidth conformation information can be supplied throughout the chain.

- *Linking the schedule with the broadcast guide.* It is useful or even a requirement for the end-customer to have access to a broadcast guide that provides overviews of the planned transmissions. A data broadcast platform should ensure that such information can be generated automatically by linking the schedule with transmission of broadcast guide information through to the end-user.

- *Co-ordinating domain names.* The nature of the Internet is that the customer's computer/device will look for content in the local cache before going out to look for it at the Internet. This means that if the content covered by a given URL already is cached, then it will pick it up from the local hard-drive and consequently not try to find it on the Internet. However, a broadcast system, while using URLs as pointers to content, will provide this content through another system other than the Internet. This means that if the broadcaster uses a URL that is used by someone on the Internet, then the broadcaster has a possibility of broadcasting different content under that same URL – and consequently to disrupt the Internet domain name system's function. A data broadcasting network must consequently devise a system for issuing URLs that do not conflict with each other, nor with the domain name system of the Internet.

6.6 BILLING

Billing is obviously an integral part of any data broadcasting platform if you want to run it as a business. However, the question is which services you need to be able to bill whom for. The main requirements are here:

- *Billing for content.* The ability to generate bills relating to the number of users within different user groups who have been entitled to receive specific best-effort services or who have actually downloaded specific packages with guaranteed package delivery. The billing capability may here be able to distinguish between which country the user is in, which category of user it is (professional/non-professional), at which time of day the user has accessed a given service, etc.

- *Billing for bandwidth.* The system should also be able to bill each sender for the bandwidth that he has used. Different channels may here be associated with specific IP multicast addresses, and services to content providers. This combination would let the network group all of the IP traffic of one – and only one – content provider onto one IP multicast address. This will mean that by measuring the IP traffic on a given multicast address, the network can get the correct information for billing the content provider behind this address. This billing should again be able to distinguish between classes of bandwidth quality, how early the bandwidth was booked, discounts offered to the specific sender, time of the day, day of the week, throughput in terms of megabytes, bandwidth, etc.

6.7 REPORTING

Content providers are often – with a good reason – very sensitive about the way that their content is being used when distributed. This is the case irrespective of whether we talk about a professional media company or a corporation, which is distributing internal material to its employees and partners/customers. A data broadcasting platform should therefore support the following reporting requirements:

- Schedule information for pending broadcasts

- Information about executed best-effort broadcasts

- Information about packages downloaded though guaranteed package delivery

- Subscription information about end-users

It should be possible to generate reports covering different aspects, such as registrations, channels, services, licences, and historical data.

6.8 MEDIA OBJECT TRACKING

If you enter a newspaper editing room, you will find that the editors have a large task in tracking their content and managing the processes whereby they work together to create and edit the newspaper. It may, for instance, work like this:

- A journalist writes a draft of a story.

- This is then submitted to an editor, who changes a bit here and there.

- He will then submit it to another editor, who re-formats it so that it fits into the page.

- A chief editor may look at the page or the paper in its entirety and then change some of the headlines.

- The final story is also filed in an electronic directory for later recall. A number of search words are attached.

All of this could easily end up as a huge mess. The common solution is to use editing software that tracks what these people are doing with the content they are creating. Without such software it would, for instance, be highly likely that the final paper would contain wrong versions of some of the stories, that some of the stories would appear twice, and that the electronic directory would miss stories, or contain versions of the stories that were not actually carried in the paper.

Data broadcasting requires similar management/tracking software for broadcast content. We may call this a "Media Object Tracking System" (MOTS). The purpose of such a system is to be able to manage the work-flow environment that supports different users and procedures as the content is being created and managed. It is, in other words, a tool that ensures that the content provider is not creating confusion and misunderstandings while working with the content.

Content for data broadcasting is obviously stored/delivered in electronic format (e.g. files) and can be of any type. The MOTS itself should not treat any of that content directly (it should not be confused with a content authoring tool), but it should support any kind of standard editing/content authoring tools used in the content creation process (e.g. HTML editors, word processors, graphic editors).

Any content that is inserted in the MOTS should be attached to a descriptor that we may call "Content Meta Data". This meta data should be used as a way of tracking the content from the point of Content Acquisition (the Content Provider), through Content Management, to Content Packaging, and thus from its most raw stage to the stage where it is ready for broadcast. Content must, through such a process, undergo a number of transformations, i.e. it is a part of the workflow process of data broadcasting, where content is created, edited, modified, aggregated, and formatted, by a number of different operators/editors. What the content meta data should do is to describe each part of the content in terms of:

- The subject it addresses

- Its position (where it is)

- Its status with respect to the workflow (how far it is in the workflow processes)

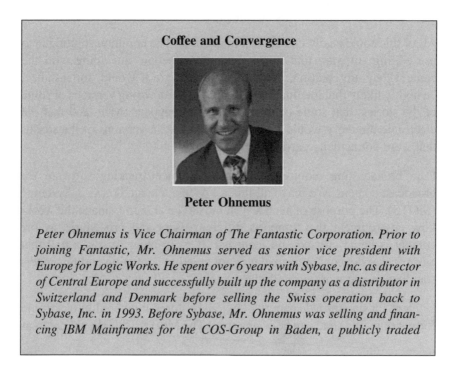

Coffee and Convergence

Peter Ohnemus

Peter Ohnemus is Vice Chairman of The Fantastic Corporation. Prior to joining Fantastic, Mr. Ohnemus served as senior vice president with Europe for Logic Works. He spent over 6 years with Sybase, Inc. as director of Central Europe and successfully built up the company as a distributor in Switzerland and Denmark before selling the Swiss operation back to Sybase, Inc. in 1993. Before Sybase, Mr. Ohnemus was selling and financing IBM Mainframes for the COS-Group in Baden, a publicly traded company on the Zürich Stock Exchange. Mr. Ohnemus is also a board member of Rochild's, Bank, Teledenanare Switzerland, Commence One and Remate-i.

One of the most interesting business meetings I have ever taken part in took place in a house on a Swiss mountainside in January 1996. We were sitting in the kitchen, sipping warm coffee, looking over the lakes and the snow-covered mountains – and sketching up business ideas. There were three people present: Frank Ewald, Lars Tvede and myself. Our discussion topic was how the Internet was evolving and where it should go, and our conclusion was that digital broadcasting and the Internet would sooner or later converge. This, we concluded, was a major opportunity.

Lars and Frank founded The Fantastic Corporation the following week, and I was able to join later in the year. We recruited a number of leading software, telecommunication and media experts and commenced the development of a series of software applications (now patented), which could be used for broadcasting virtually any kind of digital content to virtually any device with an IP address. Many of the features we developed were inspired by the way you operate digital television and telephone networks, but most of the software was based on Internet models.

The first version of the software was released in the spring of 1998, it turned out to be an immediate success.

Business today is an interdependent global economy where knowledge is the key asset and information is the currency. Two-way Internet communication is key to how you do business in the future, but so is digital broadcast; the two will live together. Data broadcasting will also change mass market entertainment. Instead of zapping from TV channel to TV channel you will be able to zap into layers of a given channel – dig in behind the pure broadcast surface to play with related content – or to take a trip into the Internet and then back to the broadcast surface – without leaving your theme. We will be better informed, work more efficiently – and have more fun.

There is still work to do before data broadcasting becomes a part of our everyday life, but it is interesting to see that leading broadband players world-wide are now preparing for it – with a full focus and with clear ideas about the applications that they want to roll out. Data broadcasting provides the needed bridge between broadcasting and the Internet and brings new services that will revolutionise the way we work, learn and get entertained.

Peter Ohnemus

Generally speaking the workflow status should be given by the "snapshot" of the values of its attributes, and the path that the content must follow should be expressed in terms of successive workflow states.

6.8.1 Digital Watermark/Copyright Handling

One special feature that a MOTS may have is "digital watermarking". This requires an explanation.

The advantages of using digital media are that they allow easy reproduction, transmission, storage and editing. These possibilities include the ability to embed data (information) within audio, image and video files. This, however, raises a copyright issue.

The solution is digital "watermarking", whereby information is hidden in the sense that it is perceptually and statistically undetectable. With many schemes, the hidden information can still be recovered if the host signal is compressed, edited or converted from digital to analogue format and back. Invisible data embedding is also referred to as:

* "Watermarking", or

* "Data hiding", or

* "Fingerprinting".

Many of the inherent advantages of digital signals increase problems associated with copyright enforcement. For this reason, creators and distributors of digital data are hesitant to provide access to their intellectual property. Copyright information can be hidden in an audio, image or video file by watermarking the file, i.e. making small *imperceptible* modifications to its samples. Unlike encryption, watermarking does not restrict access to digital media. Further, a watermark stays with the media after decryption. It is intended to provide a solid proof of ownership.

An interesting usage concerns downloading of web pages with rich multimedia content. Personalised video delivery allows users to watch a movie, broadcast to all viewers over a single channel, in a particular rating or in a given language of their choice. In this case, data embedding is used to embed extra scenes and multilingual tracks in the given version of the movie that is broadcast. Each user then extracts the hidden information that is needed to reconstruct the movie in the rating and language that he or she selected. In a sense, data embedding then provides some of the capability of DVD in a broadcast environment without requiring significant extra bandwidth or storage requirements.

6.9 INTELLIGENT CONTENT COMPILING

Data broadcasting is a dynamic medium that requires automated content processing. A critical tool here is what you might call the Intelligent Channel Compiler (ICC). ICC is an expression for a tool that handles the following processes:

- Automatic compilation of different types of data files (text, photo, graphics, video, audio)

- Conversion of these into the pre-defined formats for the data broadcasting service

- Matching of them with their pre-defined templates

- Preparing them for automatic broadcast

The type of input is first identified and then formatted according to pre-defined specifications. A search and matching process takes place next, where the software searches for corresponding multimedia data files (sound, photo, graphics, video), matches them (e.g. Word match), notes their properties (size and format) and brings the accepted matched files to the next module.

Based on content analysis and the structure of the channel, templates are pre-designed to accommodate different types or categories of content (e.g. short story template, template for particular section, etc.).

After the data is formatted, the files to be compiled are automatically selected, and the ICC reformats the content (i.e. raw text and photographs to be used, based on coding), selects the correct templates to be used, and automatically broadcasts the compiled story or presentation.

6.10 OVERALL PLATFORM FLEXIBILITY

We have now discussed a number of basic requirements for data broadcast platforms. Each of these functions enables players in the value chain to operate a data broadcast system in a commercially viable way. We shall now round off the discussion by looking at desirable features that are very general: hardware independence, network independence and browser independence.

6.10.1 Hardware Independence

Imagine that you are a content provider and that you are told that you have to create different versions of your content for each of a large number of different hardware devices. One possible reaction would be that you would choose to ignore many of these devices.

There is, however, nothing inherent in any of the concepts of data broadcasting that we have discussed previously that prevents a platform from supporting multiple hardware devices – from the PC to set top boxes, mobile devices and utilities. One issue will remain, though. Some hardware devices, like a PC, will have considerably higher storage and data processing capabilities than others, and this can mean that while a content provider produces a service for multiple devices it may not be all elements of this service that will run well on all the devices.

6.10.2 Network Independence

A content provider may also request that the content plays out well on all networks, from satellite to cable, digital terrestrial, xDSL and mobile. This is if anything easier to accomplish than hardware independence, but again there may be differences in bandwidth capacity to consider.

6.10.3 Browser Independence

Browser independence is the easiest of the three ''independencies'' to achieve. It is basically secured if the data broadcasting management software on the client side is located over the operating system (typically Winsock) but below the browser level.

7. Data Broadcasting: the Media Opportunities

"First we thought the PC was a calculator. Then we found out how to turn numbers into letters with ASCII – and we thought it was a typewriter. Then we discovered graphics, and we thought it was a television. With the World Wide Web, we've realised it's a brochure."

Douglas Adams

We have now looked at the technologies behind data broadcasting, and we have seen that it can be used as a technical work-horse that moves data around in the background. However, we have also seen that it can be used as a new medium in its own right. So let's now take a closer look at specific media possibilities that it offers.

7.1 DIFFERENT KINDS OF MEDIA EXPERIENCES

Any individual will consume electronic media in a range of very different ways, depending on the situation. These situations can be divided into 13 different main categories:

- *Background media.* Broadband content is often consumed as a background medium. The worker on the factory floor will have the radio turned on during the day, the family may have the television turned on without watching all the time, etc. The consumption of background media is completely passive, however. It is passive not necessarily because you are a couch potato, but because you are largely consuming it while doing other things.

- *Live drama.* This is the situation where you follow a live event *as it unfolds.* It might be a bicycle race, the stock market, or a breaking news story, etc. What you are interested in here is to know what is happening. You want information and you are interested in any tool that makes it possible for you to examine this information from different angles as it unfolds.

- *Community.* You are participating in a forum of like-minded people, like a chat-forum on the Internet. The forum can be social, professional, or both.

- *Game.* You are playing a game where you react to constant challenges. These challenges may be created by software/hardware technology or by other players in the game. Internet gaming, CD-ROM games and games consoles are examples.

- *Exploration.* You are interested in a subject and you search through electronic databases to find out more about it. This spans from teletext to Internet surfing and browsing to professional business information databases.

- *Contact.* You want to communicate directly with a specific individual. The typical approach here is e-mail.

- *Alert/surveillance.* You are very interested in some specific events, and you want to be alerted immediately if and when they happen. These may be sports events, or specific share prices going past a defined level. One of the most common examples is so-called electronic ''limit minders'' that investors and stock brokers set on their ''dealer screens''. They may, for instance, define a critical price for a given security, and the system will alert them if that price level is broken. They may also track their total portfolio value in a spreadsheet or portfolio management software and then be alerted if its value falls below a give threshold.

- *Analysis.* You want to analyse what some data mean. You may, for instance, track financial information in a system that continuously calculates suggested derived options pricing.

- *Learning.* You may use an electronic communication system to launch a new product for your sales people or to give technical training to students at different universities. Alternatively, a company may set up electronic billboards at public places and broadcast different promotional information to them during the day.

- *Decision.* You want to decide which colours you want in the interior of the new car that you might buy, and you are using a kiosk application or the Internet to check how different combinations would look. Or you stand at the reception of a hotel and look at local weather and traffic forecasts as well as events updates before deciding how to spend your day.

- *Emotion.* You watch a funny, scary or erotic film on television, or you see a direct broadcast from a live concert with your favourite band.

- *Transaction.* You want to buy a plane ticket to Bermuda or a ticket for a rock concert, or some books and records, computer equipment or flowers, and you choose to do it over the Internet.

- *Guidance.* You are standing in a museum listening to a taped description of the painting in front of you. Or you are lost in the centre of Milan and use your Global Positioning Manager system for direction.

When we consider this multitude of purposes that a medium can serve, then it becomes clear that you cannot generalise too much about whether a medium is ''good'' or ''bad''. Any given medium may be good for some purposes and useless for others. And so it is important to see data broadcasting for the medium it really is. Like any medium it has some strong points and some weak points, and it is of course essential to use it particularly for what it is good at. Tables 7.1 and 7.2 provide overviews of the main generic media benefits that data broadcasting can provide. The first group of benefits stems from the fact that broadcasting is totally scalable. So the first major generic advantage of data broadcasting is scalability.

The second is *speed.*

Table 7.1   Advantages of data broadcast's scalability

Unique technological benefit	Unique media benefit	Examples
Data broadcasting is scalable. The technological concept of data broadcasting means that each media object can reach any number of users simultaneously even though it is only transmitted once. This extreme scalability makes it commercially feasible to keep the network connection constantly open to each user – even at very high bandwidth. The same concept is known from common radio and television	*Suitable for "live" content.* Since data broadcasting does not have any data congestion problems it is capable of transmitting at high speeds in guaranteed real time	Live stock quotes Corporate live training video where remote viewers can e-mail questions via standard ISP connection which are addressed directly by the speaker Corporate management Lists of stolen credit card numbers
	Suitable for being "on for hours". As data broadcasting does not require a dial-up procedure or incur ISP/telco per-minute fees, it is suitable for content that the users prefer to monitor continuously, either actively or as a background medium, like radio	News and stock price monitoring Live camera views for corporate security surveillance Embedded music Emergency notification systems

Table 7.1 (*continued*)

Unique technological benefit	Unique media benefit	Examples
	Suitable for content that is needed often – by many. Data broadcasting is, due to its concept of shared bandwidth, suitable for delivering content that is required often and by a large audience	News services Weather services Corporate price lists, inventory lists, manuals, and training packages Financial information Computer games Video-enhanced interactive language training Corporate software synchronisation, such as distribution of business information databases In-store television at gasoline stations, in shopping malls, etc.

What data broadcasting *cannot* provide is the ability to connect any individual anywhere with any other individual (which is one of the main attractions of the Internet). Data broadcasting could thus never replace the Internet, but it can enrich it and relieve it of its congestion problems just as well as it can enrich traditional broadcasting, by transitioning it from single-medium distribution to multi-media distribution. And then, of course, it enables the two media flows to be accessed and manipulated within a single, seamless environment.

7.2 USERS OF THE DATA BROADCASTING MEDIUM

Data broadcasting can mainly be used to create the following five categories of media applications:

Table 7.2   Advantages of data broadcast's speed

Unique technological benefit	Unique media benefit	Examples
Data broadcasting is fast. Data broadcasting allows you to provide broadband into offices and homes without deploying a costly interactive fibre infrastructure connecting to huge play-out centres	*Suitable for content that is updated.* As data broadcasting is suitable for "always-on" connection, it is a strong solution for keeping the user abreast of events that unfold continuously	News updates Live stock quotes Weather updates Sports updates
Furthermore, it allows use of Quality of Service that secures real-time delivery	*Suitable for high-volume content.* Since data broadcast can deliver content at several megabits per second to each user, it can be used to deliver any kind of content that is too large to be suitable for traditional Internet download	Channels with embedded digital video/audio Routing of media-rich e-mails Backbone distribution of commonly used websites to ISP hubs Distribution of business software Distribution of computer games Delivery of multimedia annual reports Video-enhanced distance learning Synchronisation of large databases

- Intranet applications

- Extranet applications

- Focused affinity networks

- Branded business networks

- Branded consumer channels

which are outlined in Table 7.3.

Let us take a closer look at what each of those target groups can use data broadcasting for. We can start with the professional applications.

7.2.1 Professional Applications (Table 7.4)

There are, as we have just seen, four categories of professional data broadcast applications:

Table 7.3   Data broadcasting: application categories

Main application category	Sub application category	Explanation	Example
Professional applications	Intranet applications	Networks created by corporations or public institutions which use their internal information (perhaps combined with external information) to communicate to their staff, retailers, partners or customers	An internal system for document sharing and database synchronisation for a bank and its subsidiary network
	Extranet applications	Networks created by corporations or data broadcasting from a company/institution to its retailers, partners customers or other associated parties connected to the organisation or public institution	A distance learning system for retailers of a car manufacturer

Table 7.3 (*continued*)

Main application category	Sub application category	Explanation	Example
	Focused Affinity Networks (FANs)	Broadcasting to a network of individuals or businesses that have a shared affinity but are not initially directly connected with the broadcaster	A product information and distance learning network for doctors
	Branded business networks	Multimedia channels with business information provided by professional media vendors and offered in the open market to customers for professional use	A financial information system for any active stock market investor
Consumer applications	Branded consumer channels	These are multi-media channels provided by professional media vendors and offered in the open market to customers for private use	A games channel for any consumer interested in consumer games

- Intranets

- Extranets

- Focused Affinity Networks

- Branded business channels

Intranets are vital tools for sharing information, education and corporation vision in large organisations.

The typical concept of data broadcasting to an intranet involves broadcast to a local server for redistribution through a LAN (Figure 7.1).

Advantages of using data broadcasting as a part of a corporate Intranet include:

- *Lowers Internet access/telecommunications costs.* A single transmission is used to reach any number or groups of end users at the same time at no incremental cost.

- *Improves training.* Interactive applications are especially effective for corporate training.

Table 7.4 Data broadcasting: corporate applications

Application	Examples	Benefit provided through data broadcasting
Document distribution	Distribution of inventory lists, price lists, catalogues, multimedia staff magazines, multimedia newsletters, new product documentation, etc.	It ensures that all target users have the same version of the documents at the same time – and in time Saves distribution costs Possibility of using rich multimedia
Software distribution and updates to every PC	Distribution of software such as new browsers, application software, virus checkers, patches, etc.	It will save installation and software distribution time A data broadcast solution also helps to enforce a SOE (Standard Operating Environment)
Database synchronisation	Many corporations distribute (traditionally via FTP transfer or via mailing of CD ROMs) very large amounts of data to synchronise databases. This can instead be done via data broadcasting	Data broadcasting will save costs and ensure rapid or constant synchronisation at noticeably lower costs

Table 7.4 (*continued*)

Application	Examples	Benefit provided through data broadcasting
The "corporate academy"	Broadcasting interactive learning videos of new products and marketing strategies. The broadcast may be live and thus give employees the possibility of sending e-mail questions allowing the presenter to read and respond while "on-air". Broadcasting training videos could instruct employees of the changes in business processes and help increase acceptance of the changes	Guaranteed User Info-Levelling and Delivery (GUILD): everyone is on the same agenda because they will receive the same information Time and cost saving in educating employees By incorporating a centralised location, companies can control and track which corporate training materials are viewed, and also who viewed them and when. This allows the company to monitor corporate training for specific individuals
Crisis management	The ability to provide direct communication in crisis situations. This may include video transmission from the crisis area, interviews with key personnel, etc.	Preventing insecurity, misunderstandings and miscommunication Delivering clear instructions for action

Table 7.4 (*continued*)

Application	Examples	Benefit provided through data broadcasting
Major re-engineering projects	Data broadcasting can also be used to enhance major re-engineering projects like implementation of SAP in the corporation	A data broadcasting channel for re-engineering projects can be deployed quickly, giving the system integrator an immediate communication tool ensuring a smooth transition
Distance learning	Electronic distance learning combining video, interactive content and Internet return path	Eliminates the user's transportation problem as they can attend/access the education scenes from remote locations The transmission speed of data broadcasting allows it to transmit very rich education material and live video
Kiosk applications	Applications for displaying information such as product promotion at kiosks installed in public places such as bank affiliates, airports, libraries, hotels and shops	Operates rich kiosk applications that can be updated remotely at a low cost

Table 7.4 (*continued*)

Application	Examples	Benefit provided through data broadcasting
Market surveillance	Constant download of mission-critical business information such as business news, stock quotes, market data, exhibition news, etc. Market data is constantly updated and users can build in alerts that are triggered as critical news occurs or critical market prices are quoted	Keeping people constantly updated with edited quality data presented in a compelling environment

The frequent, rapid changes in technology ("The Economy of Speed") have increased the need for corporate training. However, the challenge

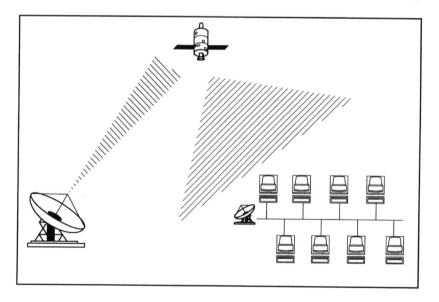

Figure 7.1 Data broadcasting for an Intranet

for corporations is not only to provide the training material, but to do it in such a way that it will capture and retain employees' attention. Data broadcasting can provide this with graphic, interactive and entertaining company-related content. The rich, multimedia content makes learning and education enjoyable, ensuring lessons are more interesting and memorable. The content could be specially produced multimedia, but it may also be a sales manager who is clicking through a PowerPoint (PowerPoint is a trademark of Microsoft) presentation that he made at the weekend while explaining it. Presenters' voices and images are broadcast as they act, and no traditional multimedia "production" is involved.

Another corporate solution is the satellite *extranet*, which connects the organisation or institution to its so-called "meta-organisation". A meta-organisation consists of people who are not direct employees, but who have an attachment to the organisation through contracts or habits. They may, for instance, be a company's retailers, its partners or its customers. An extranet application can be used as a means to communicate more efficiently with the meta-organisation.

The third group of corporate solutions is so-called "*Focused Affinity Networks*" (FANs). A FAN is a network of people or entities that have a shared affinity but are not initially directly connected with the broadcaster. The following are a few examples of possible Focused Affinity Networks:

- Distance learning network for doctors

- Financial information network for financial executives

- Information and entertainment network for farmers

- Download and sales training network for general software distributors

- Business information network for executives in the oil business

- Tracking and booking network for executives in the transportation business

The last category of professional applications is *branded business channels*. These can typically be offered by major media brands for distribution over multiple mass-market networks. The most promising subject areas are for:

- Professionals across many industries who have a shared information requirement

- Professionals who require constant search for information and/or training

Sales and marketing, finance and information technology are three areas that meet these requirements.

7.2.1.1 Securing the Corporate User's Attention

In launching a new medium, consideration must be given not only to how to obtain critical mass within the target market, but also how to obtain critical "mind-space" for each single target user. A potential problem is that people will not buy, subscribe to, install or turn-on something which has not reached critical mass of compelling content *for that particular individual*. Imagine, for instance, that an organisation has decided to launch a distance learning/information distribution network for doctors. However, they have programming available only for about 30 min/day. In time how would they avoid the doctors losing interest? There are two solutions to this problem:

- *Combine the primary content with complementary third-party content to create a more compelling end-user experience.* The doctors' network may, for instance, be enriched with pharmaceutical news, general news, weather information, financial market updates, and background music.

- *Make the channel into a targeted, vertical"Internet portal".* We have already seen how television and radio over time launched channels that were more and more targeted for specific market segments. The same can happen with Internet portals. The data broadcast channel can thus be enriched with Internet links that are particularly valuable for the target group in question. This will ensure that every time they wish to go to the Internet (those of them who do), they will routinely access it via their targeted data broadcast channel. The links within can easily be updated via broadcast. (Broadcast of a link is in most countries considered a statement of a fact, not a content broadcast transaction. This means that it is not necessary to obtain permission from the website owner in order to broadcast the links to their website.)

The result of both initiatives can be increased use of the multimedia channel.

Data broadcasting for corporate communication

Imagine that your company has subsidiaries and retailers throughout Europe. You want to introduce them to a new product, and you want to be sure that the message goes out now and is not diluted

What happens at the server end in the corporate headquarters?	What happens at the end-user end in remote offices and retailers?
Ten minutes into the presentation the VP Sales asks all the sales people to begin transmitting their questions to him via the Internet – straight away. He transmits an Internet mail addressed to himself to everybody	The Internet message "Send questions about our new product XX to VP International Sales" appears on the screens of all sales people. Some of the sales people fill it in and hit "transmit"
The VP Sales is still "on air", and he is now reading some of the e-mails on his PC in front of him and answering the questions on air	Sales staff see some of their typical questions being answered personally by the VP Sales
The VP Sales now asks all sales staff to read the PowerPoint presentation that he has forwarded and then fill in the enclosed multiple choice exam	Sales go through the PowerPoint presentation and once again listen to the embedded voice comments. They then click on the multiple choice exam, fill in the answers and transmit to headquarters via the Internet
His last remark: "Tonight we will make a draw between everybody who has returned the multiple choice form to me and filled it in correctly. The winner, who will be announced tomorrow, will get a weekend for two in Paris"	

This example illustrates the power of combining broadband broadcast (Satellite Intranet) with a traditional narrowband return path (traditional corporate Wide Area Network).

Bulls, Bears and Broadcast

Bridge™ offers a data broadcast finance channel that combines the best of the Internet with the best of television. The channel provides updates of financial information around the clock. The user selects an overview section or a more specialised section and can then watch as the financial news, market reports and key index values and price quotes for this particular area are updating. The update plays out with coverage of one item at a time, as is known from television. However, unlike television it provides the user the option to take control. It is, for instance, possible to click on "pause" to focus the coverage on a specific item, which then keeps updating. A click on "play" brings the user back to a carousel presentation. It is also possible to navigate as on the Internet, from subject to subject, but without experiencing the slow downloads of the Internet. This is achieved because data broadcasting in effect turns the user's hard drive into an Internet proxy server: the content you click for is already sitting on your hard drive.

7.2.2 Consumer Applications

It is arguably a larger task to aggregate compelling bouquets of data broadcasting applications for consumers than it is to launch a typical business solution. However, the possible consumer applications are fascinating, as shown in Table 7.5.

7.2.2.1 Securing the Consumer's Attention

What can be done from a media perspective to ensure that a data broadcast network and its content holds the consumer's attention? There are a number of possibilities to consider:

- *Create a great content mix.* There are two main dimensions to the ideal content mix. Firstly there needs to be a good combination of international and local content. A second major consideration is to combine what is a ''killer application'' with the same market segments with what are considered basic services for virtually every-

Table 7.5   Data broadcasting: consumer applications

Application	Examples	Benefit provided through data broadcasting
Enriched television Digital television signal is transmitted simultaneously with related interactive content	A TV programme for children is from time to time replaced with interactive games relating to what the children have just seen Live sports television in a part of the screen is enriched with information about timings, classifications, location of players on track/field, speed of ball, profiles of players who just scored/won, etc. News television is enriched with links to text-based in-depth explanations Travel television is using object-oriented video. Users can from time to time click on areas on the video surface to access relevant travelling information	Securing consumer intimacy by encouraging them to stay within a given channel and interacting with its interactive content

Table 7.5 (*continued*)

Application	Examples	Benefit provided through data broadcasting
Fast Internet	Transmitting "best-of-the-web" to Internet Service Provider proxies Transmitting "best-of- the-web" to end-users, thereby making their hard-drives into proxies	Speeding up the Internet Providing people who would otherwise not have access to the Internet access to limited (but very fast) content Providing an edited Internet for children
Auto-triggered Internet	Allowing transmission of digital music to trigger broadcast of a link into an Internet music-shopping site. The link provides direct information about the track, artist and album Allowing broadcast of text based news to trigger broadcast of Internet links into book shopping. The link provides direct information about books concerning the relevant subject	Providing richer information without having to create all the content yourself Providing a seamless transition between broadcasting and the Internet

Table 7.5 (*continued*)

Application	Examples	Benefit provided through data broadcasting
	Allowing transmission of business news to trigger an Internet web search with relevant information. The search profiles are broadcast to the end-users as attachments to the news stories.	
Software and games distribution	Distribute a section of a game or a software package for free use as a "try-before-you-buy" feature Distribute a complete game or software as a "try-before-you-buy" feature. The software self-destructs after a few days unless you buy the key Sell and distribute software to subscription paying customers or on a pay-per-download basis	Saving time from going to a software retail shop to purchase software Being able to try out software before buying, even if it is much too large to download via the Internet Subscribing to software, which is delivered automatically when released

Table 7.5 (*continued*)

Application	Examples	Benefit provided through data broadcasting
Advertising and e-commerce	Compelling full-screen video ads that link to electronic shopping malls. You can fill out purchase forms off-line and submit via the Internet later Last-minute travel offers appear on the screen spontaneously. You notice them since you have the system on most of the time Video actions with bidding possibility via the Internet Classified ads are broadcast. Users can put triggers in so that they are alerted instantly when the ad for the product with the right price that they were waiting for arrives	Combines the possibility of delivering compelling messages through video and CD quality audio with the possibility to search for more information, order on-line, and in some cases, get the product delivered on-line
Live gaming	Race in a virtual Formula One racing car against the real cars. The positions of the real cars are the actual ones in a real race	Adding engagement and excitement to real entertainment events

Table 7.5 (*continued*)

Application	Examples	Benefit provided through data broadcasting
Possibility to play a game in an environment that is connected to real live events	Replay goals in the breaks of a football match. The path and speed of the ball during real attempts on the goal are digitally stored, but you are now the virtual goal keeper Be a virtual stock trader. Trade virtual stocks live based on the real news and prices Bet on sports. Watch in real time how your odds of winning change as the games progress Submit your rating of speakers at a political panel while the video discussion takes place. Call up the live polled results of all submitted ratings in a part of the screen as the discussion progresses	Learning by simulating participation in complex situations as they unfold
Community	Broadcast live from the events in Internet newsgroups	Community experiences are much richer

Table 7.5 (*continued*)

Application	Examples	Benefit provided through data broadcasting
Use the broadcasting capability to enhance or create unique community experiences	People can hold e-mail interviews with a sports star. The sports star can read the e-mails as they come in and can respond to them in live video	
	Consumers can move around in a 3D virtual world on the Internet. However, while the position of each participant is controlled via the Internet, the overall environment, including rich background sounds, is transmitted to everyone via data broadcast	
	Broadcast of a multi-player game where each user can interact on-line or off-line	

one. Basic services may include news, weather, finance and TV information. Killer applications for various segments could be sports, games, sophisticated financial services, travel services, etc.

- *Provide a compelling broadcast guide.* Provide a good electronic broadcast guide, designed so that anyone can understand it, but also so that the sophisticated user is offered smart ways to select the most relevant content to watch or interact with.

- *Include a network information channel.* Deliver a communal channel to provide non-stop information concerning all channels. This could also have background music and a disc-jockey who from time to time informs the users about important events and downloads available news on each of the channels.

- *Provide good profiling opportunities.* Ensure that each user has quick and easy ways of profiling what information they wish stored on their hard drive, avoiding wasting space on excessive content that they are not interested in.

- *Make the network into an Internet portal.* Enriching the channel with links to the Internet will allow it to become the users' natural gateway into the Internet, when they wish to go beyond the broadcast content.

- *Schedule the transmission in a clever way.* Ensure content is transmitted at times when people really need it and/or have their receiving device turned on.

Creating Advertisements for Data Broadcast Channels

A data broadcast environment offers advertisers the power to produce ads which contain high quality video and audio, full-screen animation, or rich interactive images without worrying about download times, browser compatibility or plug-in conflicts.

Once the bandwidth barrier has been removed, creative opportunities present themselves, from the familiar:

- TV-like full screen video ads
- Radio-like CD quality audio ads
- Magazine-like high resolution image/text ads.

to the more adventurous:

- Hyper-banner ads driven by advanced Java scripting for enhanced interactivity
- Background audio jingles during system use
- Fly-bys (small product placement appears and disappears whilst using system)
- Rich click-throughs (detailed multimedia product information or instantly installable software demos available following click through, without download delays)
- Full screen video or animation stings between channels or features within a channel.

The areas within a data broadcast system, which could be used for advertising, include:

- Multimedia screensaversDesktop wallpaper or active desktop control
- Multimedia-enhanced direct e-mail
- Broadcast guide
- System start screen

8.   Creating Data Broadcasting Applications and Services: 26 Steps

"Every really new idea looks crazy at first."

Alfred North Whitehead

We have now discussed the basic requirements for a data broadcasting platform. The next issue is to consider its applications. A data broadcast application or service can be assembled from combinations of proprietary and open building blocks (Table 8.1).

There are similarities between creating different kinds of content, like writing a book, producing a movie, running a television news desk, authoring a website, or programming a computer game. But the differences are greater than the similarities: each medium has its own soul, with various individual approaches to consider.

Data broadcasting is first and foremost a streaming medium, which is suitable for content that is updated constantly – like radio or television. There is, however, a major difference: developing a data broadcast application or service is to a very large extent a software project. It is thus recommended that classic software development management models wholly inspire development procedures.

8.1 CREATION PROCESS OVERVIEW

We will, in the following, go through an overview of the basic steps that may be involved in creating a data broadcast application. We will, in the analysis of the process, be inspired by television and refer to such an application as a "channel". This means that we will take the concept

Table 8.1   Proprietary and open data broadcast building blocks

	Software	Media	Combined software and media
Proprietary. This means purpose-built for or by the creator of a specific data broadcast service	*Proprietary software.* This is tailor-made software for aggregating and receiving/ manipulating the content provider's content	*Proprietary content.* This is content that the content provider (corporation/ institution or media company) provides to its own application but does not offer in "broadcast ready" format for use in third-party channels	*Branded channels.* These are broadcast channels designed to provide a complete media experience for end-users
Open. This means offered by an enabler for licensing to any creator of data broadcast services	*Templates.* This is ready-made end-to-end software modules that are designed with the aim of enabling the broadcast of any content as long as this is confined within a specific format	*Media modules.* This is third party content that fits into the client interface of a channel, but which is either static or updated in a very slow cycle	*Broadcast objects.* These are end-to-end solutions containing updating content delivered within a functional end-to-end broadcast environment. Broadcast objects are licensed from third parties and used as building blocks for a channel

of a continuous channel, similar to a kind of enriched television, as the basis.

Figure 8.1 shows 26 typical steps in the creation of a data broadcast application (here a multimedia channel). We will on the following pages explain each of the 26 steps. They fall into four major phases: "planning", "building", "launching" and "operating". Going through the entire process may require from a few man-months to several man-years, depending on the purpose and complexity of the project. Table 8.2 explains which steps are required for different situations and provides a rough guideline for the manpower required.

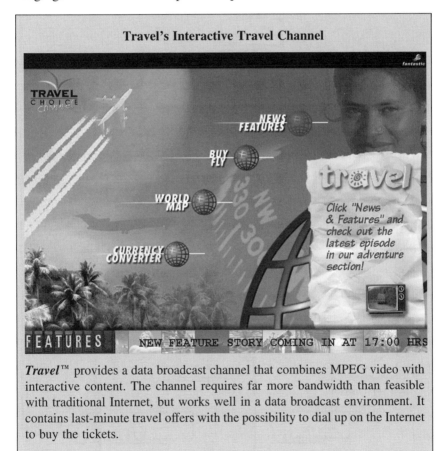

Travel™ provides a data broadcast channel that combines MPEG video with interactive content. The channel requires far more bandwidth than feasible with traditional Internet, but works well in a data broadcast environment. It contains last-minute travel offers with the possibility to dial up on the Internet to buy the tickets.

The total implementation time for a channel may, as Table 8.2 shows, range from a few months to a year, depending on the phases required and whether elements can be developed in parallel. Taking only the highest

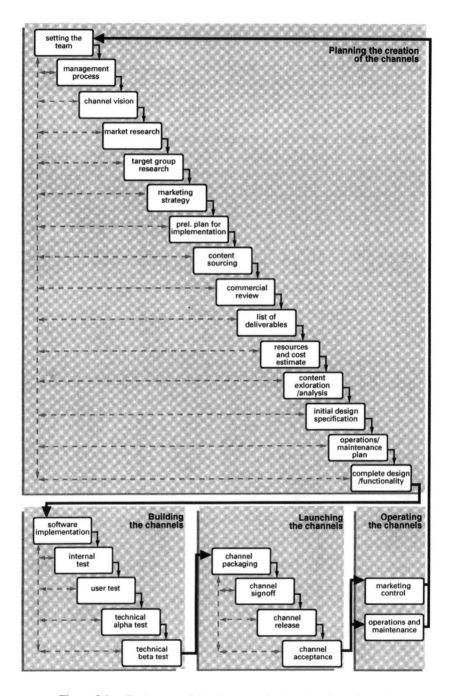

Figure 8.1.   Basic steps of development of a data broadcast channel

Table 8.2 Typical resource requirements for channel development

Main phase	Task	Typical duration of task (man-weeks)
Planning the creation of the channel	Setting the team	1
It is possible to move up and down in this part of the planning process in order to ''iterate'' towards the ideal solution	Determining the management process	1
	Channel vision	2
	Market research	4–8
	Target group research	1–4
	Marketing strategy	1–2
	Preliminary implementation plan	1–2
	Content sourcing and value proposition	1–10
	Commercial review	1
	Lists of deliverables	1
	Resources and cost estimate	1
	Content exploration and analysis	1–10
	Initial design specification	2–4
	Complete design and functionality specification	2–4
	Operations/maintenance plan	1–2
Building the channel The channel has now been described in an approved document, and the specification is frozen so that structured	Software development and implementation	1–20
	Internal test	2
	User test	1
	Technical Alpha test	
	Technical Beta test	3
	Channel packaging	1

Table 8.2 (*continued*)

Main phase	Task	Typical duration of task (man-weeks)
implementation can commence		
Launching the channel The channel has been produced and needs now to be turned into a product that is accepted by retailers or corporate administrators	Channel sign-off Channel release	1 n.a.
	Channel acceptance	2
Total development time requirements	Total	32–83

estimates from each development phase in the above table, a total of 83 man-weeks is required for developing a channel. Taking only the low estimates for a corporate channel equates to 32 weeks. However, it should be noted that these estimates are just that, and the tasks may be considerably smaller or larger relative to the scope and depth of the channel.

8.2 PROCESS DETAILS OF THE CHANNEL PLANNING PHASE

Let's now take a closer look at each of the individual steps in the process that we just reviewed. The first part is the ''planning phase''.

The detailed and skilful planning of the project is, as any experienced software developer will know, the key to success. It starts by setting the team.

8.2.1 Setting the Team

The first step in the process is to define the team for implementation of the channel. This may for a corporate channel involve internal resources only. If the channel is developed in co-operation between several entities (aggregator, system integrator, etc.), then the team should have representatives from all of these organisations.

The team will typically have personnel responsible for:

- *Product management.* Overall vision for the channel. Oversees relationships with internal/external content sources and drives the vision and commercial aspects of the channel. Writes the initial brief, which includes content, basic navigation issues, basic creative brief, and broadcast programming.

- *Producing.* Oversees the creative/technical implementation.

- *Project management.* Responsible for resource planning.

- *Multimedia effects.* Developing all multimedia effects (using standard software authoring tools).

- *Video editing.* Creating video effects.

- *3D animation.* Creating 3D effects.

- *Graphical design.* Responsible for graphical look-and-feel.

- *Intelligent channel compiling.* Creating automated content handling solutions.

- *Testing.* Technical and commercial/market testing.

- *Release management.* Overseeing the release procedures.

- *Programming/editing.* Responsible for deciding when what goes ''on air'' after launch.

- *Operation.* Managing the content flow over the broadband networks.

In addition to the team doing the hands-on implementation it is not unusual for a large project to have a reference group that functions similar to a ''board'', making overall reviews from time to time to check that the project is basically on track.

8.2.2 Determining the Management Process

The management process determines communication and decision routines for the project. It should include:

- Lists of required documentation throughout the project

- Lists of expected review and milestone meetings

- Overviews of approval processes and responsibilities

- Overviews of responsible persons for each element in the project

- Specification of release process and customer acceptance criteria for the final channel

8.2.3 Defining the Vision for the Channel

The vision of a channel is the basic, often intuitive concept for the channel. The vision for a corporate channel may, for instance, describe:

- How employees can improve their education level with the channel

- How corporate community feeling and vision sharing can be improved via the channel

- How the channel can be used to handle corporate emergencies better

- How communication costs may be reduced through use of the channel

- How speed of product roll-out and marketing may be improved through the channel

- How direct communication with end-customers can be achieved with the channel

It may be defined within:

- A geographical area

- A demographic segment

- An employee segment

- A retailer segment

- A customer segment

For a branded channel, the vision may describe:

- A demand which the channel will be particularly good at satisfying

- A customer segment which the channel will address

- A usage situation which the channel will function within

- Purchase decisions which the channel will stimulate

  and it may be defined within:

- A geographical area

- A demographic segment

- A pricing segment

8.2.4 Market Research

Market research is relevant mainly for channels that are offered to a Focused Affinity Network or for branded channels that are offered to the public. The main scope of research is to identify:

- Who should be targeted?

- Which benefits should be delivered and at which prices?

- How should the product be directed to the target groups?

- How should the target groups be made aware of the benefits that can be delivered?

One of the main preparatory tasks is here to identify the most successful medium in the market within the relevant subject area (''successful'' in terms of both popularity among end-users and also commercial success for the suppliers). Let's assume that the scope is music. What would work (and would not work) for music:

- on television?

- on radio?

- on the Web?

- in music shops?

- on other media?

This investigation will also lead to identification of the main media provi-

Table 8.3   Relevant research methods for identifying the current
                   market supply within the channel subject area

Research method	Scenario	Typical time requirements
Key interviews	You require information concerning market structures and suppliers within the channel's subject area, and you require qualified opinions and advice in addition to facts. The research may include identifying the companies having a progressive attitude to the market, the most relevant media assets, the relevant copyright structures and concerns, etc.	Varies
Distribution research	You require evaluation of relevant opinions and attitudes of executives in the distribution	Varies
Supply mapping	You need to register which electronic media products are available within the relevant market segments. This may include which products and services, in which quantities, at what prices, how they are distributed and promoted. The result can be an impression of distribution patterns, competition, price structures, and the parameters necessary to succeed in the market	A few days

ders in the business. Some of these may be obvious candidates to work
with for the channel (Table 8.3).

8.2.4.1 Identifying the Target Groups

Studies of product markets tend to show that consumers/business users have a tendency to be grouped together in clusters (segments) of similar attitudes and habits. The greater your understanding of which clusters the channel will aim for, the greater your ability to target your approach as strategically as possible. A good description of your end-users will not only provide average profiles, but also identify clusters. Identifying benefits is crucial to the success of the channel.

Understanding the benefits your product provides allows you to develop and promote it efficiently. It can also be useful to divide product benefits into the following categories:

- *Standard benefits.* These are channel characteristics providing objective benefits. This could be speedy download of software packages, non-stop access to real time, more flexible weather services, etc.

- *Corporate benefits.* These are advantages in addition to the product itself, which are brought by the supplier to the customer. They may also be the ability to upgrade to a premium version without the need for proprietary equipment, or the ability to access some of the best music titles in the business, etc.

- *Differentiation benefits.* These are advantages making the product measurably better than competing products: perhaps reduced or no dial-up hassle, automatic alerts, richer presentation, freedom from unethical content, etc.

8.2.5 End-User Segmentation

The next step is to divide main target groups into segments. End-user segmentation is a method of dividing the market for the channel into groups (segments), each of which requires a different approach, which may be in product, price, distribution and/or promotion. The segmentation may be based on criteria such as those shown in Table 8.4.

Benefit segmentation has a key role to play as it is based directly on the reasons why people use the channel.

8.2.6 Target Group Research

Once the target group has been identified it may be relevant to do some further, more detailed target group research. It is important that the functions you build into the channel answer the real needs of the identified

Table 8.4   End-user segmentation for broadcast channels

	Focused Affinity Networks and branded corporate channels	Branded consumer channels
Geography	Countries Region	Countries Region
Demographics	Economic status Culture/ethnic areas Political orientation	Economic status Culture/ethnic groups Political orientation
Social/sector criteria	Corporate professional level Business sector Size of the company/organisation Production facilities Financial status	Education level Profession Household income Lifestyle group
Personal criteria	End-user's gender End-user's position in the organisation End-user's age	Gender Age Marital status Size of household (number of people)
Price sensitivity	Price sensitivity	Price sensitivity
Purchasing situation for e-Commerce/pay-per-event applications	Size and frequency of ordering Importance of the purchase Brand loyalty Geographical purchase preferences Selection criteria in the purchase situation	Usage intensity Importance of the product to the user Brand loyalty Geographical purchase preferences Selection criteria in the purchase situation

Table 8.4 (*continued*)

	Focused Affinity Networks and branded corporate channels	Branded consumer channels
Product usage	How is the channel used (always on, or driven by Programme Guide Announcements, mainly at what time of the day?) Which features in the channel are most appreciated?	How is the channel used (always on, or driven by Programme Guide Announcements, mainly at what time of the day?) Which features in the channel are most appreciated? What are the channel service/support requirements?
	What are the channel service/support requirements? What are the channel substitution possibilities?	What are the channel substitution possibilities?
Benefits	Standard benefits Corporate benefits Differentiation benefits	Standard benefits Corporate benefits Differentiation benefits

target groups, whether they are employees, retailers, Focused Affinity Groups, corporations at large, or segments of consumers.

Assuming that the target group is *internal* (intranet/extranet solutions), then they themselves will be ''customers'', and the channel will have to be ''sold'' to them. The best way to do this is to give them something they need and cannot get elsewhere. To determine what this could be, an analysis is necessary. A good method is to start to gather representatives from different departments and ask them what information they would like to publish on the channel. There may be many different sources of information from within these departments. Discussions directly with the end-users are also useful. It is important to analyse how they will relate to the information or messages. Will this influence your choice of content, its structure, timing and presentation? Will the target groups or their demands change over time? How will one track these changes? Segmen-

Table 8.5   Formal target group research methods

Research method	Scenario	Typical time requirements
Focus groups	You do not yet have a structured view on respondents' attitudes to your (still) loosely formulated ideas about the channel. You may be looking to create the basis for later quantitative tests. Research is conducted by gathering a group of people (typically eight approximately) at a location where a moderator oversees the debate	Approximately 4 weeks
Informal interviews	Same situation as Focus Groups, but interviews are done on an individual/informal basis	1–4 weeks

tation is also important. Departments, management level or content categories can also be used to further segment a channel, or services within the channels. It is also possible to direct specific services (sub-channels) to individuals or groups with a Subscription Management System.

If the target groups are external (Focused Affinity Networks and branded channels), then it might be relevant to perform more formal research, as shown, for example, in Table 8.5.

8.2.7 Channel Marketing Strategy

Although there may be an internal marketing issue for corporate channels, the term ''marketing strategy'' is here meant for the commercial marketing of data broadcast channels to Focused Affinity Networks and the public at large. The purpose of the marketing strategy is to determine:

- Product policy for the channel

- Pricing policy for the channel

- Distribution policy for the channel

- Promotion policy for the channel

8.2.8 Preliminary Implementation Plan

It is vital to have a basic guideline as to how the channel should be implemented or rolled out before the channel implementation starts. For a corporate solution the following steps may be anticipated for:

- Production of a dummy-run

- Launching a simple live ''proof-of-concept'' channel to a few test locations

- Launching full regional channel

- Launching full global channel

For a mass-market channel the plan might specify a controlled expansion in a limited area in order to gain experience before moving to a full regional and then international augmentation.

8.2.9 Content Sourcing and Value Proposition

Internal content sourcing is done through three steps:

- *Identify the concept that the content of the channel should follow.* A good understanding of the target groups and desired benefits of the channel should lead to an identification of what content the channel should contain.

- *Analyse what relevant content is readily available.* Internal channels can be created with content from the company's/institution's intranet site, websites, search engine query results, news groups, files on the network and databases.

- *Analyse what relevant content could easily be created for the data broadcasting medium.* Analyse if the medium raises new opportunities, such as broadcasting live video, that have not been utilised before, simply because a suitable delivery mechanism was not available.

For external sourcing the task is to identify and meet with the content providers or Media Aggregators that can offer the most relevant solutions. An agreement with these must then be negotiated.

8.2.9.1 Information Retrieval in the Corporate Environment

Most corporations produce a massive amount of just-in-time information that would be suitable as components in the corporate data broadcasting channel. Once these sources have been identified the next task is to find a way of retrieving them automatically. Several software tools already available on the market can help structuring before publishing documents and information on a broadband channel. Examples of software that might be used are Channel Manager, Wincite, and Intraspect.

8.2.9.2 Copyright Issues

It is always necessary and in some cases not trivial to ensure that all content used in the channel is owned by the supplier. In cases where external content is used (photographs, video, graphics, text, etc.), the chain of ownership should be fully determined and contractual right for use obtained from all parties in the chain.

If you are unable to make a full check of the chain of rights, ask the content source for an explicit guarantee clause in their authorisation contract. The contract must comply with certain rules, if the right of a third party for use is to be validly assigned.

In addition, ensure that all employees involved in the creation of the channel, including text, video, photos, sound (voice over, music), graphics, etc. have a clause in their employment contract assigning all rights to work created during employment to the employer ("talent release form").

8.2.10 Preliminary Commercial Review

The commercial review is an examination of the projected development costs, operation and media asset acquisition for the channel as well as of direct and indirect revenue forecasts. There will typically be several scenarios ranging from "Best Case" to "Worst Case".

8.2.11 Lists of Deliverables

The list of deliverables is a complete list of all the input that each party in the project has to deliver during the planning horizon.

8.2.12 Resources and Cost Estimate

The resources and cost estimate provides an overview of all expected costs of the channel project.

From an editorial point, a channel will typically consist of various elements that require different degrees of effort and cost. Table 8.6 lists some examples.

8.2.13 Content Exploration and Analysis

The content exploration and analysis phase aims at securing a full under-standing of the content that the channel software environment (and the end-user) has to be able to deal with. This can be done by:

- *Obtaining a daily sample production of the relevant content.* This may be delivered in any original electronic format that will be applied during the actual operation of the channel. It is vital to make it clear to the content providers that the format they have shown is the precise format that will be used when the channel is in operation. This is the only way to ensure that the channel operations process can be automated.

- *Obtaining a live feed of the relevant content.* Some content providers will already have a constant stream of updating data in place, and it may, in these cases, be beneficial to obtain direct access to these sources. The content provider may then have a defined API (Application Interface) available in print.

- *Obtaining protocol descriptions and other written documentation of the content.* Large media companies currently in the business of providing streams of updated electronic content may already have very professional documentation for this content, thus value resellers and other users are able to use the content in different contexts. These descriptions form a valuable basis for writing the Intelligent Channel Compiler software that automates the method in which the channel content is processed.

Table 8.6   Effort and cost overview for different content categories

Channel component	Examples	Development effort	Operations effort
Visual environment	Graphics, background and logos	A one-time effort	None
Static content	Instruction sections, disclaimers	A one-time effort	None
Semi-static content	Advertising	Requires manual effort at lower intervals	Limited
Updating content	News, features, downloads	Requires manual effort at regular intervals	Extended effort depending on level
Dynamic/real time content	Financial data, stock quotes	Automation requires development/ identification/ integration of automated feed-handler tools (Intelligent Channel Compilers)	Significant programming effort, which is often specific to each feed.

8.2.14 Initial Design Specification

The design specification defines the design rules and guidelines for the channel.

The most important elements of the specification are listed in Table 8.7.

8.2.15 Channel Operations and Maintenance Plan

The channel operations plan will explain how the daily flow of content from source to receiver is supposed to be managed. It includes the tasks shown in Table 8.8.

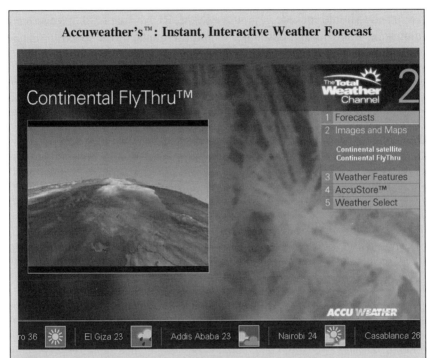

Accuweather's™: Instant, Interactive Weather Forecast

Accuweather™ has been one of the data broadcast pioneers by providing a rich, interactive weather channel. The service combines instantly available 3D fly-through weather images with satellite images, interactive weather maps, a multimedia weather education centre, e-commerce and much more. The channel provides rich, graphical weather animations as known from television with the ability to select what and when as known from the Internet. This screen image shows a weather image with a 3D fly-through animation playing. This animation is based on a 600 kbyte file, which would take time to download on the Internet. However, in Accuweather's™ service it starts playing within approximately 1 s after the user has clicked on the menu. This is possible since the file has been background downloaded to the hard drive through data broadcast.

Table 8.7   Key elements for the initial design specification

Basic item	Examples
Defining which hardware platform the channel shall be able to run on. (Definition should include minimum, standard and optimal specifications and what to expect for the channel performance in each case)	Set top box, PC, mobile device, etc. RAM, ROM, free hard drive space requirements, clock speeds Assumed communication cards, sound cards, video cards
Defining which software platform the channel shall be able to run on	Operating systems, browsers, assumed browser plug-ins, screen resolutions, display colours, etc.
Content file types, formats and attributes planned	Types: Graphic files, sound files, movie files, animation files, text files, etc. Formats: JPEG, GIF, Wave, AVI, MPEG2, Shockwave, Flash, GIFs, etc. Attributes: ''Wave is 22050 Hz 8 bit Mono, 22 kbps, etc.''
Screen size and resolution	''794 × 490'', ''780 × 565'', ''800 × 600'' or ''640 × 480'' Possibility of using tilted background
Software applications to be used	Front page Adobe Photoshop Adobe Premier Adobe Illustrator Adobe After Effects Macromedia Freehand Macromedia Flash Macromedia Director Macromedia Sound forge Macromedia Extreme3D

Table 8.7 (*continued*)

Basic item	Examples
Navigation	Navigation in relationship with the subscription management structure Rules for consistent internal navigation

Table 8.8   Channel operations: brief overview of tasks

Operations task	Explanation
Setting up communication lines for receiving content into the Media Aggregation Centre	How does the content flow from its sources into the centre where it is being aggregated (via satellite, the Internet, LAN, WAN, dedicated land-lines, mailing of CD ROMs, FTP, etc.)? How can content be inserted downstream?
Network broadcasting scheduling plan	When shall the different content components be broadcast and at which bandwidth? How often shall content be retransmitted when assuming that some of the users do not have their receiving devices turned on all the time?
Monitoring down-time incidents	How is it ensured that breakdowns in the inbound/outbound communication are immediately detected and handled?
Monitoring media incidents	How do you track incidents and handle where inbound content does not meet agreed standards (arrives late, in wrong format, etc.)?
Monitoring software incidents	How are software bugs reported and handled?

Table 8.8 (*continued*)

Operations task	Explanation
Managing incidents and requests for development	How are incident reports and requests for development of the channel communicated and handled?
Delivery from content from Media Aggregation Centre to Network Operation Centre	How is the aggregated content delivered from the Media Aggregation Centre to the network Operation Centre? (via satellite, ADSL, LAN, WAN, dedicated landlines, etc.)
Monitoring content flow to user groups through Media Object Tracking Systems	How can you monitor how content flows to each of the defined user groups? (for internal reasons and/or for being able to report to content providers)
Operating Subscription Management System	How are different users enabled/disabled for each of the services in the channel?

8.2.16 Complete Design and Functionality Specification

This is the plan that finally "freezes" the complete description of the channel. The main sections of the specification may, for instance, be:

- Introduction

- Project organisation

- Management process

- Channel vision

- Target group analysis

- The usage scenario

- Channel marketing strategy

- Channel deployment plan

- Content exploration and analysis

- Design specification

- Operations plan

- Channel structure

- Graphical elements

- Formats

8.3 PROCESS DETAILS OF THE CHANNEL IMPLEMENTATION PHASE

We have now looked at the steps of *planning* the channel development. The next step is to actually *implement* the channel. While every effort may have been put into planning of the channel it is still possible that issues are discovered making it necessary to modify minor elements of the plan. It is vital that any changes are:

- Tracked with new improved revisions added into the plan

- Checked against schedule and cost estimates

- Formally approved by the project team

The first Channel Implementation phase is software development (Table 8.9).

Once the software has been developed the channel is still a long way from being finished. The process of building channels can be complex since it is necessary to work with many files, all linked together. It is consequently vital not to lose control over the structure of interaction between different files and not to lose track of the interim module versions (can be merged with off-the-shelf versioning and control software).

After software development you get to the testing stage. Before you release the channel for broadcast it must be tested. It is advisable that incremental testing is effected as you build the channel in order to detect errors progressively, in the early stages. These incremental tests are called ''internal tests''.

Table 8.9   Main software elements of a data broadcast channel

Main software development component	Explanation
Intelligent Content Compiler	Software which takes care of retrieving, validating, reformatting, integrating and re-transmitting the content of the channel
Media Object Tracking System	Software that tracks implementation of changes in the transmission settings as well as the target groups for each content transmission
Client interface	A browser-based application that organises the storage, display and manipulation of the content on the client application

8.3.1 Internal Test

The internal test follows the completion of the software development. The first internal test of the channel can be performed in six steps:

- "Basic sanity test": test whether the channel can be loaded and its content accessed within the prescribed environment.

- Short term test of the server and the client software on single, isolated platforms.

- Test transmission of updating content from server through a network emulator to client device.

- Test transmission of updating content from server through a network emulator to several client devices with different configurations.

- Test transmission of updating content to client device, which is simultaneously receiving other data broadcasting channels or other content.

- Test transmission of updating content through a real network environment.

The internal test should serve not only to identify bugs, but also to identify functionality issues. These can be mapped by following a check-list:

- Is the text readable?

- Check all text for spelling errors and incorrect formatting.

- Do all the pages load in an acceptable time frame?

- Do all the sounds play at the right times and at the correct volume?

- Check video and animation files for playing correctly and in the right places.

- Do the users like the colours and graphics in the channel?

- Is the sound a good accompaniment or is it annoying and repetitive?

- Is the level of information in the channel interesting and useful?

After these tests, the channel (or parts of it, e.g. services) is ready from a design and content aspect to be taken for external user test.

8.3.2 User Test

The purpose of this testing is to discover if the channel works as planned when confronting the unprepared user. Is the channel understandable? Can the user navigate around the channel easily? Is the content complete? You can present the testers with a list of tasks to perform, which could include searching for the answer to a specific question requiring them to navigate through the channel. Ensure they are provided with a feedback form. It can also be useful to videotape the users while they are trying to use the channel. Experience has shown that uninitiated users tend to detect bugs that skilled computer professionals do not find. The test may lead to discovery of bugs as well as any cumbersome/inconvenient devices that can be handled through minor (or major) graphical design modifications, better tutorials, etc.

8.3.2.1 User Tests for a Data Broadcast Channel During the Implementation Process

There are various efficient ways to test a data broadcast channel as it is being conceptualised and implemented so that the development can be adjusted to the market requirements. The most useful methods are listed below.

Test	Explanation	Typical duration from test decision until a report is available
Concept tests	An evaluation of the channel concept is required in order to determine if it is likely to work, what its strongest and weakest elements are and – if there are several alternatives – which one is best. The test may be conducted individually for the channel and for its promotion. The test can for instance be effected with a macromedia mock-up of how the channel may appear at implementation	Approx. 4 weeks
Product tests	This test is quite similar to the concept test, but is focused on specific features rather than on the general channel concept	Approx. 6 weeks
Concept/ product test	An evaluation of the channel concepts and the actual channel features is required in order to evaluate the overall impression and possible discrepancies between concept and product	Approx. 8 weeks

Each of the tests will take less time if conducted with internal staff.

8.3.3 Technical Alpha Test

The alpha test is the first duration test run in the internal environment.

Testers will follow a test plan that will include various stress tests and random clicking by uninitiated personnel.

8.3.4 Technical Beta Test

The beta test will take place among real users. These users can be employees or "friendly" external users. Beta testers should of course be made fully aware that the channel is not commercially released.

8.4 PROCESS DETAILS OF THE CHANNEL LAUNCH PHASE

Once the channel has been through a successful beta test it will be time to launch it commercially. This involves packaging, internal sign-off, release and customer acceptance.

8.4.1 Channel Packaging

The channel must still be productised, or "packaged" before launch. The typical steps within this process are to create at least some of the following components:

- Installation programs and disks for the client software.

- User guides for distributors (and for end-users) if the channel is complex.

- Electronic channel tutorials.

- Channel product sheets including hardware/software specifications for the end-users.

- Channel rate cards for retailers.

- Promotion material for retailers, advertisers, end-users, etc.

8.4.2 Channel Sign-off

The channel sign-off is the internal process where everybody working with the channel agrees that it has reached a stage and a quality that makes it possible to operate professionally. A sign-off may use a check list for critical issues.

8.4.3 Channel Release

The process of releasing a data broadcast channel is largely a communication task. The users need to be informed that it is no longer a vision but readily available for use. This process can be aided by anything from mass marketing, press announcements/internal announcements to presentation meetings, distribution of demos, etc.

8.4.4 Channel Acceptance

The acceptance test is prevalent as the data broadcast channel is offered as a commercial product to third party aggregators or retailers.

A data broadcast channel is an electronic product with a significant software component, and thus can be expected to go through a customer acceptance test much like other software products. The acceptance criteria for such a test may be stipulated within the channel's distribution contract. It is in any case normal that implicit silent acceptance is achieved if and when the retailer offers the channel to the first end-customer.

9. Operating Networks of Multiple Data Broadcasting Channels

"If computers get too powerful, we can organise them into a committee – that will do them in."

Bradley's Bromide

There is a difference between the task of *creating* a data broadcasting channel and the task of *operating a network* of data broadcasting channels. A channel creation project is largely focused on software issues with a beginning and an end. Operation, on the other hand, will in principle never end.

Channel operation is in many ways reminiscent of managing the content of a multi-channel television network. It is continuous, and it involves monitoring a number of different variables, which raises a number of editorial, technical and commercial issues (Table 9.1). Each of these issues is considered below.

9.1 EDITORIAL ISSUES FOR MULTI-CHANNEL NETWORK OPERATION

9.1.1 Multi-Channel Scheduling

A major issue to consider when operating many channels simultaneously is how to optimise the combined content scheduling. Basic criteria to consider for this planning are:

- *When are the users expected to have their receiving devices turned on?* If you transmit to a user who has not got his receiving device turned on, he will not see what is being transmitted, and the device will not store it for delayed viewing. It is necessary therefore to transmit relevant content at times when the relevant users are most likely to have their receiving devices turned on. While many corporate users will have their receiving devices (typically PCs) turned on

Table 9.1　　Channel operational issues after the release

Editorial issues	Technical issues	Commercial issues
Multi-channel scheduling	Providing a basic channel operations infrastructure	Product policy
Broadcast guide and information channel operations	Data gathering and processing procedures	Support infrastructure
Electronic tutorial	Backbone transportation network	Media object tracking
Test channel	Data on-ramping	Subscription management
	Client software deployment Incident/R+D handling	Network marketing control

all the time, this is hardly a realistic assumption for consumers as a PC is, after all, noisy, heat generating and power consuming.

- *When are the end-users for each broadcast object expected to view the content?* If, for instance, you are transmitting programming for children at home, then it is best to start the transmission when they come home from school, and continue until they go to bed. Distance learning for universities could be transmitted during the day, financial prices when the relevant exchanges are open, etc.

- *How critical is it to back up the content?* For an example, assume that you are an active private investor. You are consequently subscribing to a financial channel that contains updated charts (graphs) with stock prices. You go on a 2-week holiday, and your PC is meanwhile turned off. Problem: when you come back you find that there are large gaps in all the charts. The data from your 2-week vacation is missing. This is an example of data that needs to be backed up through clever retransmission schemes, with an option to dial in via the Internet if the broadcast back-up still leaves gaps.

- *What is the ideal balance between use of bandwidth and use of hard drive space for the client?* The best example of this issue is the transmission of video-clips. Let's assume that a 90-s video-clip takes up 15 Mbytes hard drive. This is not a problem on its own, but what if you want to show 20 of those clips during a day? You can either deliver them as files that are stored on the hard drive, or you can stream them through, in which case they are not stored. If you choose to stream them through, you save disk space; however, you might want to stream each of them through several times during the day, as you do not know when target users will be using the system.

- *Are there time zone differences among the users?* The ideal time for transmitting a given programme package depends on the time zone you transmit to. If you have 5 h time difference between two audience groups, then you might have to transmit twice as often as you would for a single group.

- *What is the total bandwidth available?* It is necessary to fit each of the individual scheduling plans into an overall scheduling plan, thus enabling total bandwidth not to exceed the available network capacity.

- *Do bandwidth reservations for data spikes/peaks leave frequent spare bandwidth that can be used for transmission for back-up data?* Assume that you have reserved bandwidth for a news channel that allows live MPEG streaming whenever there is a crucial "breaking news" story. However, when there is not breaking news video transmission, only text news and pictures are being transmitted from time to time. The transmission will thus vary within a range of 10–128 kbps most of the time, spiking at 2 Mbps some of the time. So why not use all of that spare bandwidth to retransmit data? It is possible to set the system to automatically fill up the dedicated bandwidth with retransmissions of data from this or another channel.

9.1.2 Broadcast Guide and Information Channel Operations

Another operational scheduling issue is the broadcast guide. It is essential when operating data broadcasting channel networks to ensure that users receive proper updated information concerning the content that will be "aired" (the equivalent being television guides, which appear as listings within common newspapers and magazines, on Teletext, on the Internet, as electronic programme guides on digital television set top boxes or in numerous glossy TV guide magazines). Maintaining a good Broadcast Guide for a data broadcast based network is particularly important since data broadcasting provides a combination of live streaming content and content that is cached/saved for later use. For example, should there be a download of a free section from the very latest high-tech game, when will it take place? At what alternative times will the gamers need to keep their PCs/set top boxes on in order to receive it?

The scheduling software tool should ideally create the Broadcast Guide information automatically. Whenever a transmission is scheduled the system should prompt you to give it a name and then automatically place a representation of it in the Broadcast Guide, which is transmitting this information to all clients at regular intervals.

However, it is not optimal to leave it at that. Some of the content will need to be detailed for the end clients, inserting explanatory text, video and sound files in some places to give the end-user a better idea of what to expect. This means maintaining regular contact with the content providers, whether they are media companies, media aggregators or corporate departments, in order to understand better what should be communicated about the programming.

An additional method of communicating the programming is to run a dedicated information channel. This concept, which is familiar to cable television companies, can be developed to communicate the programming events to a far greater extent. Such a channel may, for instance, provide streaming music interrupted from time to time with information about key downloads and live events on the network. The information could include anything like:

- ''Those of you who are interested in financial information should turn to the financial channel. The Dow Jones Index has just made a large jump, and there are some exciting news stories to explain it''

- ''Have you heard about the new version of xxxx browser? You can receive a free trial only this week. We will be speed downloading it to all users during the first 3 min of every hour of the day – around the clock. The campaign runs until Sunday evening at midnight''

- ''Calling all cool kids: 3 min from now we are downloading the funniest game of all time. In only a few whopper-snapping minutes you will be having the Komplete Kidorama experience. Get ready on the Kids Channel and check it out''

- ''The new sports channel is here. Click on the main menu to see the new button. That is how you get to it. You can watch video with interactive background information beginning tonight. The first 3 months are free!''

9.1.3 Electronic Tutorial

The electronic tutorial should provide a short introduction to how the channel works. It should not take more than a few minutes to go through the tutorial. Some tips for making it:

- It may use elements from the information channel in order to develop a coherent brand identity.

- It should show examples on how you navigate the system. There should be voice-overs explaining these examples.

- There should be a particular explanation of how the test channel and the broadcast guide works.

- It should be clear to the user that broadcasts only provide content when the receiving device is turned on.

- The more complex reference sections should be separated for the advanced user to consult.

9.1.4 Test Channel

It is advisable to run a test channel that end-users can use for diagnostics of the system in order to check that they are receiving content. The channel should retransmit some simple content changing constantly in a carousel.

9.2 TECHNICAL ISSUES FOR MULTI-CHANNEL OPERATION

9.2.1 Providing a Basic Channel Operations Infrastructure

One of the first technical priorities in channel operations is to ensure there is a clear separation between each of the following technical environments:

- Corporate infrastructure environment

- Channel development environment

- Channel testing environment

- Channel demonstration environment

- Channel operations environment

Within the channel operations environment the support of safe transmission 24 h a day, 7 days a week must be assured. The following elements need particular consideration:

- Around-the-clock surveillance

- Automatic alarms if any data feeds stops transmitting

- System redundancy

- Basic security, such as power back-ups, antimagnetic environment, etc.

It is particularly vital for reliable operations that full redundancy on all critical hardware components is secured in case of any breakdown. Duplication of satellite down-links and server hardware is thus essential. The system should ideally switch automatically from one unit to its back-up in the event of a breakdown. Alternatively it might sound an alarm, in order for an operator to make the switch manually before diagnosing the error.

9.2.2 Data Gathering and Processing Procedures

The original content is required to flow automatically from the content sources (which may, for instance, be media companies for branded channels or corporate departments for intranets/extranets) into the Channel Operations Centre. The transportation into the centre can be effected via many alternative means, such as satellite download, dedicated landline, FTP, the Internet or even physical distribution of tapes, CD-ROMs, etc. Electronic delivery is of course preferable as it makes it possible to automate the data handling completely.

9.2.3 Backbone Transportation Network

The purpose of a backbone transportation network is:

- To transmit the content from its central aggregation centres to the Network Operation Centres (NOCs) of the retailing networks/corporations,

- In some cases to interconnect Media Aggregation Centres (relevant if one Media Aggregator runs several Media Aggregation Centres).

It is more practical in almost all instances to use satellites for this network as satellites have the ability to reach any number of NOCs within their footprint areas. Some of the typical considerations in deployment of these networks are:

- *Satellite footprints.* Which areas are covered with which transmission power (at transponder transmission saturation or normal operational power levels if lower)?

- *Planned launch dates for satellites still not deployed.* Failures at launch are not uncommon and can delay a satellite deployment considerably.

- *Expected remaining lifetime in case of existing satellites.* Satellites have limited lifetime (typically 10–15 years). It is important to have plans in place for the replacement of satellites with limited remaining lifetime.

- *Precise orbital location.* This is normally quoted with ± 0.1 degrees accuracy.

- *Fixed antennae support.* The design orbital stability of the satellite must be sufficient to enable the use of fixed antennae.

- *Security.* It should be considered whether the bandwidth offered is "pre-emptable" or "non-pre-emptable". Pre-emptable means that there are no back-up solutions guaranteed by the provider if the primary solution fails. In non-pre-emptable solutions it is relevant to know the status of on-board spare capacity and/or capacity located on another space vehicle.

- *Up-link and down-link hardware specifications.* It is relevant to know which size and specifications of the up-links and down-links must be used.

- *Signal availability.* Consideration must be given to how much statistical signal downtime can be accepted from the satellite operator.

- *Band.* The satellite can transmit in CA-band (bi-directional), ku-band (uni-directional, requires small receive antennae), or C-band (uni-directional, requires small receive antennae).

- *Bit rates.* This is the bandwidth offered.

- *Licences, landing rights.* Legal permissions to receive data from, or transmit data to, satellites in the areas covered.

- *Redundancy.* Should both receive and transmit stations have back-ups?

9.2.4 Data On-Ramping

Assuming that data broadcast channels flow from an aggregation centre to a corporation or another aggregator, how would the receiver handle the "on-ramping" of additional content such as local branded channels or corporate information?

9.2.5 Client Software Deployment

The client software should typically contain:

- Basic enabling software ("browser plug-ins")

- Navigation interface to channels

- Broadcast guide

- Possibly elements of content to look at until the first transmission goes through

- Electronic tutorial

- Test channel interface

There are two basic issues to consider regarding deployment of this client software to the end-user:

- How to deliver the initial installation software

- How to follow up with software updates

9.2.6 Delivering the Initial Installation Software

The basic alternatives for delivering the initial installation are shown in Table 9.2.

Each solution has its advantages and disadvantages:

- *Physical delivery.* One advantage is that with physical delivery on a CD-ROM/DVD the user sees instantly the access to the system. It is, in other words, a kind of communication tool. A further advantage is that you are assured that the users actually receive the software. A disadvantage is that the content of the CD-ROM/DVD may be dated when it reaches the end-user.

- *Electronic delivery.* The major advantage is the flexibility. Whatever you broadcast is the latest version. A disadvantage is that you cannot

Table 9.2 Software installation alternatives

	Stand-alone	**Embedded**
Physical	Each end-user receives a CD-ROM with the client software interface	PCs and other devices are sold with the client interface CD included, pre-installed CD-ROMs with the client software are distributed with standard computer magazines, etc.
Electronic	The interface is broadcast to the client upon taking out a subscription	PCs and other devices are sold with the client interface CD included, pre-installed The interface is broadcast in conjunction with traditional television signals The interface is downloaded via the Internet

be certain whether users are connected when the broadcast takes place. Further, in order to ensure that the software is received correctly you have to use package delivery, which requires an Internet return path for all users.

- *Stand-alone.* The advantage of this solution is in the independence of the habits of the user. A disadvantage is that it requires the user to have actively committed to use the system and know how to install from a CD-ROM/DVD.

- *Embedded.* The main advantage is that it is an efficient way to sell the concept. Furthermore, in the case of pre-installation on PVC you have the advantage that the user does not have to take any actions at all to make the software work.

9.2.7 Following Up with Software Upgrades

Whenever there are new core software releases, new channels, new interfaces within the channels, or as subscribers change subscription configuration the issue is raised as to how to deliver the software changes to the relevant clients.

The most efficient solution is in most cases to broadcast all changes (using package delivery) on a regular cycle in the expectation that the user sooner or later will have the receive device turned on during a transmission. However, it is still necessary to have a contingency solution. Assume that the core software goes though a number of upgrades and that some channels are not fully backwards compatible. How do you ensure that these users discover that the reason their content is not displayed is that they have not received the relevant core software as yet?

One solution is to program in such a way that the users with a dated navigation interface automatically see a notification in the channel stating the problem and its solution (when the core software will be downloaded next time, or how a back-up delivery of it may take place). Another is to use a package delivery solution, so that missing parts can be retrieved via the Internet.

9.2.8 Incident/RFD Handling

It is necessary to implement an infrastructure whereby all incidents and requests for development are recorded and acted upon.

9.3 COMMERCIAL ISSUES FOR MULTI-CHANNEL OPERATION

9.3.1 Product Policy for the Channel Network

When offering a network of channels it is necessary to consider the following key questions:

- Channel portfolio strategy

- Channel bundling strategy

- Channel roll-out flow

- Channel linking strategy

9.3.1.1 Channel portfolio Strategy

A sound channel portfolio will typically contain a combination of three types of channels: "killer applications", "must-haves", and "nice-to-haves". Table 9.3 shows some examples of what this might be.

9.3.1.2 Channel Bundling Strategy

Channel bundling strategies are very well known in the television world, where you typically can choose a basic package and upgrade packages. The same strategy can be applied for branded channels and FANs, where for instance you might choose to offer:

- Basic package

- Silver package

- Gold package

A basic package may be low cost or even free-to-air (i.e. free if received from satellite). The same channel will typically have a price when received via another infrastructure such as a cable network. Other packages may target different user groups altogether. There might also be individual premium channels that can be accessed individually.

9.3.1.3 Channel Roll-Out Flow

It might be useful to launch channels gradually for the users to experience the excitement of a gradual improvement. Further, consideration should

Table 9.3   Framework for a balanced channel portfolio

	Killer applications	Must-haves	Nice-to-haves
Explanation	Applications that by themselves can motivate some people to subscribe	Applications that you expect to find, even if they are not by themselves enough to justify subscription	Applications improving the perceived quality of the network, but which are not considered vital
Examples from intranet/extranet/ FAN solutions	Dedicated distance learning	General business news	Background music option
Examples from branded business channels	Targeted business news Real time stock prices On-line language training	Weather General business news Weather channel	General news Back-ground music option General news
Examples from consumer channels	Games Sports Erotic entertainment Teen lifestyle entertainment Home finance Science and adventure	General news Weather Travel Health and fitness	Background music option Food and drink Culture

be given to launching free-to-air initially in order to build up a reference base of customers.

9.3.1.4 Channel Linking Strategy

There are a number of questions relating to the combination/linking of content to consider:

- *Which combination of channels is needed in order to make strong multi-channel packages?* A consumer offering might for instance include a combination of "killer-applications" and "must-haves".

- *How should the channels be bundled?* Should each channel be offered on the same basis or should there be packages/bundles? It might, for instance, be considered to offer a basic package free-to-air with a silver package and a gold package in two different pricing ranges.

- *Who should deliver the content?* Should the channel be based on internal media assets only, or should external ones be included?

- *Should components of the channels be marketed to other broadcast channel operators?* It may be appropriate to develop off-the-shelf "Broadcast Objects" which can be sold to other channel operators (a Broadcast Object can be defined as a live feed of specific broadband content delivered within a productised end-to-end broadcast software environment).

- *Should the channels contain e-shopping, advertising, pay-per-event, etc?* Which revenue sources would fit with the scope of the channels and which would be the simplest to implement from a practical point of view?

- *Should the channels interface with a website or a TV channel?* The content provider may have the possibility of providing smart links and cross-media experiences between the data broadcast channel and other media assets.

- *How should the channel evolve over time?* Which channels and channel features could be added over time?

9.3.2 Support Infrastructure

It is necessary to ensure an infrastructure so that end-users can receive support whenever they have a problem.

9.3.3. Media Object Tracking

One of the key purposes of Media Object Tracking is to be able to report to content providers where their content is going, when and in what shape. This tracking and subsequent reporting must reflect deals with content providers.

9.3.4. Subscription Management

Current subscription management procedures are required for any service that contains other than free-to-air content.

What is a Broadcast Object?

A Broadcast Object can be defined as a live feed of specific broadband content delivered within a productised end-to-end broadcast software environment. It may, for instance, be a targeted news channel, a channel with financial price quotes, a channel with background MPEG music streams or a channel with weather updates that can be readily embedded into corporate Satellite Data Broadcast applications. For example, it could supplement data provided by a corporation running its own satellite data broadcast network. Such a corporation could, for instance, provide its own content to a channel for farmers, supplemented by general agriculture/food processing news from a general news provider like Reuters. In such a case, Reuters might deliver the content in the form of a ready-made Broadcast Media Object using ready-made software.

The main promise of this approach is the speed to market. Rather than having to develop the solution by itself, the company could simply call the content provider, who would then open the data stream within hours or even minutes. In addition to the data stream it would also broadcast (and retransmit) the browser ''plug-in'' needed to display the content of the Broadcast Object.

9.3.5 Channel Network Marketing Control

Marketing control is relevant for data broadcast channels that are launched to external entities. Table 9.4 indicates some of the most useful tools for ensuring the data broadcasting channel business is on track.

Table 9.4   Marketing control for data broadcast channels

Basic question	Operational control indicators (actual versus planned situation)
"Do the market and the competition behave as planned?"	Monitoring market growth and behaviour Monitoring competitor behaviour
"Does our promotion work as planned?"	Monitoring development in relative brand awareness for the channel, channel usage patterns and preference for the company and its competitors among end-users Monitoring promotion efficiency for each medium used Monitoring image and market position among Media Aggregators (if relevant) and among the channel's end-users
"Does our pricing work as planned?" "Do our sales and distribution work as planned?"	Monitoring competitive price levels at end-user level Monitoring sales volumes and market share developments per distribution network segment Monitoring sales volumes and market share developments per end-user segment Monitoring the distributor's satisfaction with products and services
"Do our products and services work as planned?"	Monitoring the end-user's satisfaction with products and services (opinion stated, resales and reference sales ratios, returned goods ratios, complaints, etc.) Monitoring channel down-time and bugs
"Do we have the planned marketing profitability?"	Monitoring cost and margin per contact and order Monitoring cost and margin per distribution and end-user segment

10. The Commercial Drivers behind Data Broadcasting

"When a fellow says it ain't the money but the principle of the thing, it's the money."

Kim Hubbard

It goes without saying that no business can thrive in the long term unless there is a clear net value creation. So where is this value creation in the data broadcasting business?

Data broadcasting creates fundamental added value in various ways. *For the professional world,* data broadcasting can create value by:

- *Improving professional education.* Ability to educate in a more effective way due to the richness of the media.

- *Improving professional information.* Ability to communicate directly to people in real time with a strong media interface to convey visions, instruct in crisis handling, provide corporate information, advertising, etc.

- *Improving professional motivation.* Conveying in a more effective way what the corporation/institution is all about.

- *Saving corporate/institutional communication costs.* Reduction of the telecommunication costs via the use of shared bandwidth.

- *Saving people's time.* Reduction in the time taken to distribute media via other methods.

- *Compressing corporate product launch cycles.* Making it possible to send strong, detailed messages from central locations to people "in the field" ensuring that corporate tasks can be implemented immediately, in a coherent way by everybody involved.

- *Saving corporate/institutional money through disintermediation.* Providing the ability to deliver very high-volume electronic products and services direct to the corporate user, reducing distribution and end-user costs.

For the consumer world, data broadcasting can create value by:

- *Improving consumer education.* Ability to educate in a better way because of the richness of the media and to provide this education to people in remote locations.

- *Improving consumer information.* Combining more entertaining media interfaces with the possibility to ''dive in'' for more detail.

- *Improving consumer entertainment.* Providing broadcast entertainment that can be interacted with, thus saving money through disintermediation. Providing the ability to deliver very high-volume electronic products and services direct to the consumer, reducing distributor and end-user costs.

- *Saving people's time.* Reduction or elimination of waiting time.

- *Stimulating literacy.* Providing interactive content to people who would not use the Internet because of technophobia, economic reasons or lack of access.

These basic values translate by various means into money, which trickles via complex configurations to the various value chain players. As the basic value translates into specific value for each value chain player it will appear in a variety of forms – some of which will have limited resemblance to the basic revenue drivers. This does not change the fact that it is the basic, fundamental value of data broadcasting to the users which is the ultimate driving force of the business.

We can here look briefly at how this flow of value through the chain is likely to evolve.

10.1 HOW VALUE FLOWS THROUGH THE VALUE CHAIN

Data broadcasting is likely to evolve in a step-by-step enabling and marketing process, until the value chain is finally complete and the real value for end-users is generated. These natural steps are:

Phase one: basic core technology enabling

- Software and hardware providers develop the core technology making data broadcasting possible and commercially feasible.

- Network operators (satellite, ADSL, cable, digital terrestrial, mobile) build or license the fundamental technology that can enable them to offer flexible bandwidth bit-stream and content services to data broadcasting clients.

Phase two: basic commercial enabling

- Media companies decide to provide content to data broadcasting services and applications.

- Media aggregators assist large media companies in building data broadcasting channels/services.

- Advertising sales agencies agree to sell advertisements for the medium.

- Advertising metering providers agree to provide advertising metering for the medium.

- E-commerce providers agree to provide e-commerce facilities through data broadcasting channels.

Phase three: early adopters

- Corporations and institutions license the enabling software and acquire bandwidth from networks in order to deploy data broadcasting solutions for intranets/extranets/FANs.

- System integrators are introduced to the technology and start integrating and reselling the core technologies for clients.

- On-line retailers start aggregating channels into broadband portals and offering them to end-customers.

Phase four: growth and maturity

- Enablers offer complete solutions rather than technology.

- The solutions are very user-friendly and non-intrusive.

- They are widely embedded into other mainstream technology.

- The solutions are offered in a range of different versions targeting different market segments.

- There is a range of data broadcasting content applications available.

- There is general convergence around common open standards.

10.2 SHARING THE ADDED VALUE THROUGHOUT THE VALUE CHAIN

We just looked at the deployment in the value chain from a time perspective. Another angle is to look at the motivations for each of the market players throughout the market phases.

10.2.1 Value for Basic Core Technology Enablers

The first activity in the market starts when entrepreneurial companies (enablers) decide to ''plunge in'' to create and sell the basic technology for data broadcasting. The main motivation may be the expectation that this market will, over time, become as big as radio and television and that its key technology users (telcos, satellite operators, large media companies, multinational corporations, cable companies, etc.) can afford to pay for quality technology solutions. Under this assumption there are obvious major opportunities for entrepreneurs to lead the creation and move early into the learning curve of creating the necessary core hardware and software. The core technologies developed by these companies will allow network operators to offer the core data broadcasting solutions: booking broadcast bandwidth on a flexible basis with the content being displayed in a way that can be handled at the client side. But why would the networks want to do that?

Because broadband bandwidth business increasingly has become a commodity business. Both senders and receivers have a range of competing broadband bandwidth connections to choose between. There are cable companies, digital terrestrial, competing satellite operators, competing telcos, and soon even mobile broadband providers. In time the decisive sales parameter for pure bandwidth will thus become *price*: if two suppliers offer essentially the same solution, then the customer is in a position to negotiate them down to the point where very little or no margin may be left. Being enabled for data broadcasting can provide three solutions to that problem:

- *Ability to segment the market.* Broadband bandwidth tends to be sold on a rather simple basis. Access to satellite bandwidth is, for instance, typically sold on long-term contracts – say 3 years. This is not a mature business model. In comparing it to how airline tickets are sold, you do not reserve a seat from New York to Frankfurt on a 3 year basis. Airlines use complex market segmentation; they have first class, business class and tourist class. There are weekend offers, holiday offers, and membership bonus points. And you pay different prices depending on whether you book well in advance or you book at the last minute. Data broadcasting uses (unlike, for instance, analogue television) highly flexible bandwidth. Some users might want to transmit nothing for several hours, then go up in the megabit range, then drop to a few hundred kilobits, etc. Some might demand guaranteed real time throughput (e.g. stock quotes), while others can tolerate limited buffering. Some might be happy to transmit overnight (e.g. database synchronisation), while others need prime time access. Some can plan their bandwidth usage well in advance while others need full flexibility. So being able to sell bandwidth with the flexibility that data broadcasting technology ideally provides makes it possible to segment the market and create a win–win situation with the customers.

- *Ability to provide value-added services.* "Value-added services" is always the strategic advantage necessary for any commodity business. Offering data broadcasting can be a solution as it may offer anyone who connects to a given broadband network access to broadcast content. Assume, for instance, that there are news, weather, finance and music data broadcast channels available on a given network. With nothing available on the competitor's network, which network will the customers prefer to use?

- *Ability to attract new customers.* The third motive is simple. It is the ability to attract data broadcasting customers in the first place. The broadband network that cannot offer the basic data broadcast services necessary will obviously not attract customers looking to broadcast multimedia.

The result of these three advantages for the network will be additional bandwidth sales and higher average prices.

10.2.2 Value for Basic Commercial Enablers

It is an old lesson that it is difficult (or impossible) to sell new core technology for electronic communication if you do not ensure that there is content for it. There are here three scenarios:

- The buyers of the technology are themselves creators of relevant content for it (example: telephones)

- The buyers of the technology are not creators of content for it (example: radios)

- Some of the technology buyers create relevant content for it, others do not (example: VCR)

Data broadcasting falls into the third category. Corporations might have the relevant content for their data broadcasting channels available in-house, but mass market service retailers will typically not. Data broadcasting will be slow to take off unless there are companies ensuring that a critical mass of publicly available content is offered early on. This will not only create a reason for users to adopt the medium, but it will also teach others how the medium can be used. This phase can be called "commercial enabling".

The commercial enabling phase involves five different market players:

- Media companies

- Media aggregators

- Advertising companies

- Advertising metering providers

- E-commerce providers

10.2.2.1 Media Companies

Media companies have one strong motivation to adopt the medium. Experience shows that each time a new core technology for electronic media appears, there are a number of smart early adopters (innovators) who move into the business fast and subsequently capture the key market segments. Good examples are BBC (first television and radio), CNN (global news television), Discovery (science and adventure television), Eurosport (European sports television), MTV (music television), BskyB

(British satellite television), Yahoo! (Internet portal) and Reuters (real time financial information). The "first-mover advantage" gained by moving in early on is particularly strong within broadcast media, where there is only room for a limited number of channels on a given infrastructure. Once domination of a segment is achieved within the new medium, it can be extremely difficult for competitors to catch up. So the motivation for media companies is to ensure that they position themselves to dominate their market segment, create brand awareness within it, and move into the learning curve of packaging content for it before their competitors, thus building a profitable business with great cross-promotion opportunities around it.

10.2.2.2 Media Aggregators

Media aggregators can perhaps best be seen as the "plumbers" of the business. They fill the following basic roles:

- Collect the updating content

- Write software to automate the handling of it (intelligent channel compilers, media object tracking systems, client interfaces)

- Process the continuous content aggregating a reformatting

- Manage the content scheduling

- Insert advertising and e-shopping facilities

- Optimise the scheduling in conjunction with scheduling of other channels

- Optimise the simultaneous handling of multiple channels on the client side

- Wholesale the channels to retailers

- Deliver the channels via backbone infrastructure to the hubs or head-ends of retailers

A good example of a similar business model is Reuters for financial information. Reuters collects data from banks, brokers and electronic stock exchanges, etc. and aggregates it through automated data handling (software processing) before it is redistributed through a number of different networks to the end-users, who view the aggregated content within a rich software environment. Media aggregators' key assets are often not their content *per se*, but their different abilities to add critical value through clever software processing and networking.

10.2.2.3 Advertising Companies

How is a decision to place an advert in given media made?

- The corporate marketing director is working with an advertising agency.

- The agency uses statistical planning software and subscribes to media usage statistics for different advertising media.

- The marketing manager and the agency agrees on an advertising budget and a communication objective for the advertising activities.

- The agency produces a media plan, showing how the company can, on average, reach people (1) within the optimum target groups, (2) an optimal number of times (say, for instance, 13 times), (3) over an optimum time interval (should be the typical realistic time to get from product awareness to product purchase), and (4) with messages that fit into the strengths and weaknesses of each medium.

- Once the plan is approved the agency will either place the ads directly with the media owners (if it is a full service agency) or it will place them through advertising sales agencies.

The problem that always needs to be overcome is how to move the potential customers through the different phases of the sales cycle. Different media have different strengths and weaknesses. Table 10.1 compares television with the Internet.

These different combinations of strengths and weaknesses mean that the majority of campaigns are forced to combine a variety of different media, which further complicates the planning. Although statistically media correlation can be checked, there will be a large number of individuals who will not be exposed to the media impressions a number of times and in the planned sales cycle order.

One of the problems here is that people move through the sales cycle at very different speeds. What if you see a television ad and want to know more straight away? Do you go into another room, dial into the Internet, type down the name of the product and browse through 11,256 potentially

Table 10.1 Television, the Internet and the sales cycle

Phase in sales cycle	Explanation of phase	Optimal tools for the phase	Strength of television as a medium	Strength of the Internet as a medium
Create customer awareness	Creating emotions via compelling and entertaining presentations	Strong, media rich ads with quality video and audio	Very strong	Weak
Enable collection of information	Enable the customer to understand the relevant benefits of the offering	Interactive catalogues where customers can examine details and alternatives	Weak	Very strong
Create customer involvement	Create positive attitude and mental involvement	Interactive presentation where customers can see examples or simulations/ calculations of personalised solutions	Weak	Strong
Allow customers to compare with alternatives	Show that this is a generally accepted concept and a competitive offer in the marketplace	Interactive catalogues where alternatives are lined up	Weak	Very strong

Table 10.1 (*continued*)

Phase in sales cycle	Explanation of phase	Optimal tools for the phase	Strength of television as a medium	Strength of the Internet as a medium
Allow customers to purchase	Ordering and paying for the product	Electronic ordering and payment (perhaps with credit card)	Weak	Strong
Deliver to customer	Get the product delivered	Electronic download of the product, if it is electronic	Weak	Weak

relevant websites? Or what if you see the ad on the television and want to buy straight away? The best solution that television occasionally offers is that you can quickly find a pen and paper, write down a telephone number that you see on a screen, then go to the telephone and order the product. This is not very practical.

Data broadcasting can overcome these difficulties as it can accommodate all the phases: from creating awareness (video ad) to enabling collection of information; creating involvement and allowing comparison (interactive catalogues); purchasing on-line (link directly from the browser interface into the Internet); and, in some cases, downloading the product (secure package delivery at several megabits per second).

The interest of the advertising companies is to understand and use this medium for more targeted and efficient advertising. Through this they are able to offer better services to their customers within a competitive environment.

10.2.2.4 Advertising Metering Providers

It is difficult to sell advertisements for a medium if no one produces reliable statistics concerning the audience of that medium and the ads. Advertisers will naturally expect accountability and comparability, which can be provided by independent advertising metering companies.

What Buyers and Sellers of Electronic Ads Require

Buyers of advertising are either the advertisers themselves, or, as in most cases, advertising agencies that are creating and managing media campaigns on behalf of advertisers. Their needs lie in four categories:

● *Creation needs*. As a creative agency creates the advertising message, they look for research to determine if their idea meets the desired objectives. For example if the purpose of a campaign is to primarily brand a new product, the desired objective is that the majority of consumers (if not all) remember and recognise the product advertised after they have seen the ad. This type of research is usually done in a focus group setting where a group or respondents are exposed to the idea and then, using aided and unaided recall interviewing techniques, the researcher tries to determine the effectiveness of the ad. Also, the researcher can determine what attributes the respondent remembers about the product advertised based upon the advertising message. For example, 200 people are shown 30 min of a TV show with a few adverts in between. The target ad (a Mercedes Benz car) is within the 30-min viewing but the respondents do not know that this is what they will be tested on. After viewing, the interviewers will administer a questionnaire to the respondents and ask them which ads they remember seeing (unaided recall). It is possible that of the 200, a few may not even remember seeing the Mercedes Benz ad. Next, the interviewer may ask specifically if the respondents remember seeing a Mercedes Benz ad. Some may still say "no" (bad creativity!). The interviewer may then ask what they remember about the ad and what they think are the key product attributes (safety, comfort, performance, prestige, reliability, etc.). The responses are then used to determine if the advertising is having the impact that the advertiser wants. A good research project has a "control cell" which is a group of people who are not shown anything yet administered a questionnaire. The responses here act as the basis and any deviation from this in the exposed setting is attributed to the advertising. For example if 10% of the control cell recalls Mercedes as a safe car and 50% of the respondents who saw the ad say that, the impact of the ad was 40%.

● *Media planning needs*. Once a marketing strategy has been decided, the next step in the process is to ensure it reaches the target audience, determined by traditional media research tools. The "Nielsen numbers", which indicate what audience is watching what shows, is, for instance often used to guide the media plan for TV. The media plan looks at placing the commercial in different shows at different times to achieve the level of desired exposure with the target audience. Also feeding in here is rate information. Clients look to come up with a media plan and mix that can deliver the desired media objectives (in terms of audience reach, frequency of reach and audience composition) at the lowest possible cost. They look at alternative outlets (ads in NBC vs. ABC may get the same audience but NBC may be cheaper to buy on a CPM – cost per thousand impression – basis than ABC) and finalise the buy.

• *Proof of performance needs.* Once a campaign is launched, the broadcasters offer various guarantees and create terms for the campaign. For example, the agency may be offered at least 1MM impressions to be delivered in 5 days. Based on historical information (see the previous point above) both the agency and the broadcaster may feel that this may be achieved by running the ad in three shows. However, it is possible that once the ads start running, actual audiences may be less or more than expected. If they are less, the broadcaster may be required to refund the money (for the undelivered impressions) or "make good" and keep running the ads until the required number of impressions are delivered. Only when the conditions of the contract are met and confirmed by audience research numbers does the agency pay the advertisers. In some last-minute ad buys (known as the "scatter market"), the broadcaster may sell unsold inventory very cheaply but offer no guarantees.Competitive advertising information. One of the critical components of marketing planning is competition awareness. Coke wants to know where, what and to whom Pepsi is advertising and vice versa. This information is collected using ad intelligence tools and the ad occurrence information is tied to known information about how much the ad cost. This allows for reporting not only on where a specific product is being advertised but also to whom and how much the advertiser is spending to reach the audience. Competitive information is vital to advertisers and agencies constantly needing to refine campaigns and re-position products and messages in the marketplace.

Sellers of advertising need proof of performance and competitive advertising information:

• *Proof of performance.* Sellers of advertising look for proof of performance that they can deliver to the buyers in order to collect the revenues earned to them.

• *Competitive advertising information.* Sellers also look at competitive advertising information to track who is spending how much and where, to assist in developing sales pitches to attract advertisers to their outlets. For example, NBC may see that Coke spends a lot on ABC and that they could offer Coke the same audience levels for less money. They would then pitch this proposal to Coke.

Table 10.2 provides an overview of how advertising measurements are handled by different media.

10.2.2.5 E-Commerce Providers

E-commerce businesses are mainly gaining ground for the same basic reasons that physical shopping malls and supermarkets originally took market share from traditional retail shops: they offer greater information, larger selection, more convenience and lower prices. An e-commerce solution tends to be competitive if it can provide better information, larger selection or more convenience.

Better information:

- Provide general news relating to the product subject area

- Provide ability to search through a database

- Offers and products that change frequently or consistently

- Provide flexible/auctioned prices

- Provide community experiences around the products

Larger selection:

- Provide visual/audio presentations of product alternatives at a distance

- Provide and display a broader product selection

More convenience:

- Provide solutions to people who are far away from the closest physical retail outlet

- Provide the convenience of having an electronic account for repeat purchases

- Provide the advantage for the customer to have electronic tracking of purchases

- Provide the ability to shop outside normal shopping hours

- The products are electronic and can be delivered via the on-line connection

The question is now: can a data broadcasting enhanced electronic shop-

Table 10.2   Media measurements for advertising

Question	TV	Internet	Data broadcasting
Creative testing	Focus group	Off-line or on-line focus group	Off-line or on-line focus group
Media Planning	Panel based historical demographic information	Panel based demographic information and site logs detailing complete site activity	Panel based demographics collected at client end. Census level tracking (no demos) is also possible, if there is a back channel to a central facility that allows for the collection of usage activity without compromising user privacy
Proof of performance	Combining audience information from panel with ad occurrence information collected usually by means of an identifying code inserted in the ad by the advertiser	Ad tracking on websites with panel information on demographic profile of audience exposed to message. However, given that the content and the advertising are not as tightly integrated as in TV (two people hitting a website at the same time may get different	Same as media planning

Table 10.2 (*continued*)

Question	TV	Internet	Data broadcasting
		ad messages), the panel based demographics does not provide as good a means to determine audience to demographics as they do in TV	
Competitive advertising	Passive systems to track ad occurrence information that do not require any code insertion or co-operation by advertiser (important to be passive as co-operation usually comes with restrictions on what can and cannot be done with data – releasing to competitors is a 'No')	Same as TV. However, estimating expenditures may be very difficult given that rates are driven by targetability and exposure of campaign; and since on the Internet, there is little relation between content exposure and ad exposure, using content exposure as a surrogate for ad exposure (as is done for TV) can be misleading	Exposures can be linked to content but actual interaction can be customised. For example, all may see the same BMW ad but some may choose to look at safety features while others may choose to look at price. Competitive ad tracking will use components of TV and Internet tracking

ping mall provide advantages over a traditional e-commerce solution? It can in many cases, as shown in Table 10.3.

Metering in a Data Broadcast Environment

Manish Bhatia

Manish Bhatia has been working with Nielsen Media Research since 1989. He is currently heading Nielsen Media's Internet research division and is responsible for the development, launch, sales and marketing of Nielsen Media's Nielsen//NetRatings service.

Manish is also working to develop measurement solutions for the convergence devices including Win98 (Microsoft), cable set-top boxes (Worldgate Communications) and the broadband distribution of interactive content on PCs (Intel Intercast, The Fantastic Corporation). Before getting involved in Interactive research, Manish was involved in developing Nielsen's upcoming Active/Passive metering system for digital TV and Nielsen's new AMOLII content encoding system. Before that, Manish worked in Nielsen's cable division (NHI) and worked on major Nielsen accounts including Turner, HBO, MTV, etc.

Manish holds an MBA in Computer Information Systems from Baruch College, City University of New York and a Masters in Economics from India. He has also completed Advanced Computer Networking courses at New York University.

Data broadcasting is a uniquely powerful environment that combines the power of TV and the Internet while eliminating key limitations of both media. It enables content producers and advertisers to reach a mass audience in a medium that closely resembles one they are very familiar with – TV – and allows them to offer enhanced and interactive experiences without requiring Internet connectivity at the user end or having to develop back end infrastructure that the Internet necessitates to fulfil interactive requests from the hundreds of thousands, if not millions, of viewers and consumers the content producers want to reach.

Measuring viewer response in this new medium requires adaptation of the existing TV and Internet measurement models to capture the user behaviours that are relevant for this medium, and in some cases, unique to this medium.

Nielsen Media Research has developed a complete measurement system that allows us to capture the full range of activities users of the system can engage in. The system is broad enough to enable the reporting of activity to meet the various needs of the content and advertising communities using this platform.

As a first step, we will be reporting on the number of people who have access to this platform. This is a critical first piece of research information that content producers and advertisers need in order to determine the total potential audience size – the total reach – made available by this new platform. Combined with the gross number will be additional information on the demographics of the persons and households having the platform. This information can be collected during the registration process the user goes through when they buy and install the platform.

On an ongoing basis, we would be tracking the actual usage of the platform. This would be done by activating a Nielsen tracking software in a randomly selected, representative panel of users who agree to participate and allow Nielsen Media to track their ongoing behaviour for research purposes. The selected and co-operating panel members would provide more detailed information about the households and individual characteristics such as age, education, income, etc. In addition, questions about product purchase history and likelihood to buy specific products can be collected. This information will be combined with the ongoing usage information collected from the co-operating panel members. This would allow us to answer various questions content producers and advertisers will have:

How much time do different people spend on viewing/interacting with content on the data broadcast platform?How much time do they spend viewing/interacting with my content/advertising message?

Who are the people that are interacting with my content? How are they different from the overall user base? Am I over/under delivering on certain types of people?

The above information can be combined with supplemental information collected during the recruitment process to understand, in greater detail, the audience being reached and the impact the message is having on the audience on the data broadcast platform.

Further, similar information from other media (TV, radio, print, etc.) can be used to do a cross-media comparison to determine how much of the total media budget should be allocated to a data broadcast platform.

Manish Bhatia

Table 10.3 Comparison of the traditional e-commerce solution with the data broadcasting solution

	Traditional e-commerce website	**E-commerce incrementally improved through data broadcasting**
Better information		
Provide related news	Can provide updating and limited multi-media about the subject area to attract audience	Can add streaming video and audio with, for instance, author interviews for books, music video clips, travel destination videos, etc.
Database search	Can provide the possibility to search through large databases for subject areas	Can be enhanced in the case of classified advertising by streaming updates through as streaming data. Users can then leave systems running and set alerts whenever there is a relevant update such as the job you might be looking for in the area where you live

Table 10.3 (*continued*)

	Traditional e-commerce website	E-commerce incrementally improved through data broadcasting
Update on frequently changing offerings	Changes will be available and updated whenever you log in to check	Updates can be tracked in graphs that update live
Communicate flexible pricing	Price developments can be illustrated graphically	Live cameras tracking auctions or rich, constantly updating graphics tracking bids, ask and volume as people trade live
Participate in product related community experiences	There can be chat forums, "people who bought this have also bought...", live top seller ratings, etc.	Broadcast live from the events in Internet newsgroups, stream live video interviews with relevant personalities and commentators; possibility to ask e-mail questions

Larger selection

Visual/audio product presentations	Pictures of product selections, like cars in alternative colour configurations, etc.	Use of streaming video and audio to provide enhanced presentations
Wider product selection	Large catalogue with more products than you could display in a physical world	Guided multimedia tours through the virtual world

Table 10.3 (*continued*)

	Traditional e-commerce website	E-commerce incrementally improved through data broadcasting
More convenience		
Bridging physical distances	Available anywhere for anyone with an ISP connection	Available anywhere for anyone with a data broadcast connection (might be satellite, cable, digital terrestrial, ADSL or mobile)
Maintaining electronic account for repeat purchases	One-click ordering	No additional benefits
Electronic tracking of purchases	Check your electronic shopping basket so that you know how much you are going to spend before submitting	No additional benefits
Shopping outside normal hours	Shop around the clock	No additional benefits
Lower prices		
Sampling or delivering electronic products on-line	Download of games, software packages, financial reports, etc.	Possibility to download very large files (such as, for instance, 100 Mbytes) through package delivery

Data broadcasting may, as illustrated above, enhance the attractiveness of an electronic shopping mall if it has:

- *News components* (e.g. book reviews for books, financial news for financial services, events for ticket sales, press presentations for new cars)

- *Classified ads components* (e.g. cars, houses, home appliances, holiday house rentals, jobs)

- *Frequently changing offerings*(e.g. last-minute travel offerings, equity prices, food wholesale, home electronics)

- *Flexible prices* (e.g. groceries, used cars, antiques, auctioned airline tickets)

- *Related community experiences* (e.g. music, books, movies, sports)

- *Need for visual/audio product presentations* (e.g. paintings, houses, travel destinations, fashion)

- *Need for a large product selection* (e.g. computer appliances, software, industrial supplies, music and book titles)

- *Customers with a physical distance problem* (e.g. software, games, travel tickets, concert tickets)

- *Products that can be delivered on-line* (e.g. games software, business software, graphical libraries, business reports, annual reports, browser plug-ins)

The last feature (ability to deliver very large electronic products on-line through package delivery and large bandwidth) is perhaps the most important added benefit. It is a clear case of disintermediation providing faster fulfilment and lower costs.

10.2.3 Value for Early Adopters

The third phase in the market starts as the solutions are deployed for end-users. This occurs as:

- Corporations and institutions license the enabling software and acquire bandwidth from networks in order to deploy data broadcasting solutions for intranets/extranets.

- On-line retailers start aggregating branded channels and offering them to end-customers.

The early adopters are organisations wishing to gain a key competitive edge and who are willing to move early in order to improve their competitive position.

10.2.3.1 Value for Corporations and Institutions

Corporate users will be motivated by fundamental value motives since they represent the end of the value chain (representing their own end-users):

- Improving professional education

- Improving professional information

- Improving professional motivation

- Saving corporate/institutional communication costs

- Saving people's time

- Compressing corporate product launch cycles

- Saving corporate/institutional money through disintermediation

10.2.3.2 Value for On-Line Retailers

On-line retailers may, unlike corporate users, have some motives differing from their clients' motives. One major motive can be to generate a so-called ''Internet Portal'' status. An Internet portal is a commonly used gateway to the Internet. The benefit here is the ability to sell advertising and charge commissions from the e-commerce sites as people enter from your portal. You are a virtual reseller. However, one problem for traditional Internet portals can arise as users after a while get so acquainted with their preferred e-commerce sites that they bypass the portal and go straight to the destination site. The portal has thus provided transitory value by promoting the e-commerce site; however, over time this value may fade away as the e-commerce sites establish their own brand loyalty. Alternatively, a data broadcast based portal will have a more sustainable position portal because:

- It can carry ads that are richer/smarter than ads in any other medium.

- It can enrich the e-commerce experience through enhanced information, selection and convenience.

- It can run always-on, meaning that there is rich media running within it even before you have dialled into the Internet. Since it runs within a browser it is destined to become the natural launch path for an Internet session.

10.2.3.3 Value for System Integrators

Corporations, institutions and on-line retailers will all have their preferred system integrators. These integrators have an interest in being able to provide or support a large range of state-of-the-art solutions to their clients. Data broadcasting contributes to that and is thus an added business opportunity for them.

10.2.4 Value in the Growth and Maturity Phase

The fourth phase starts as the market moves from early adopter stage to mainstream where now:

- Enablers offer complete solutions rather than technology

- The solutions are very user-friendly

- They are widely embedded into other mainstream technology

- The solutions are offered in a range of different versions targeting different market segments

- There is a range of content applications available

- There is general convergence around common open standards

Each of these steps will happen because they create value for different players as shown in Table 10.4.

Table 10.4   Value drivers in the mature data broadcasting market

Mature market characteristic	Value drivers
Enablers offer complete solutions rather than technology	Allows convenience and security for the customer
The solutions are very user-friendly	Reduces derived costs for the customer

Table 10.4 (*continued*)

Mature market characteristic	Value drivers
They are widely embedded into other mainstream technology	Compatibility is ensured, and the deployment is to some degree automatic
The solutions are offered in a range of different versions targeting different market segments	Each customer gets only what he needs
There is a range of content applications available	The technology can quickly be put into use for a range of different purposes
There is general convergence around common open standards	Media companies and corporations can create content with the knowledge that it can be played out on many different devices and network infrastructures. Also means lower cost of technology components and a larger pool of people educated in use of the technology

11. Data Broadcasting in the Future

"Where a calculator on the ENIAC is equipped with 18,000 vacuum tubes and weighs 30 tons, computers in the future may have only 1,000 vacuum tubes and perhaps weigh 1–11/2 tons."

Popular Mechanics, 1949

We have now looked at data broadcasting as a technology and a business – mainly from a contemporary perspective. But what about the future? Where could data broadcasting lead us in the longer term?

Firstly it should be remembered that data broadcasting is a "core technology" as it opens the gateways for an avalanche of new applications that were not possible or commercially feasible before. It does this by providing the basic capability of broadcasting, constantly updating multimedia to hundreds of millions of users through an infrastructure that is commercially viable (it is largely already in place). As a new core technology it provides a range of new opportunities geared around:

- When and where people access content

- What content people access

- Which people access the content

The future is notoriously difficult to predict, but let's just make a few guesses. We can start by considering *when and where* data broadcasting content may be received.

Data broadcasting is an extremely cost-efficient way of delivering bulky/rich digital data to large audiences. So one of the predictable implications is that many devices in the future will evolve with the ability to receive data broadcast content. Some possibilities:

- Your mobile phone is constantly receiving and displaying financial data, weather forecasts and news (receiving only does not drain the battery nearly as much as bi-directional communication such as TCP/ IP).

- Your car will constantly receive information about road conditions. If you code in where you want to go, then it will automatically alert you to road congestion on your route.

- You can set your PC to track your investment portfolio or the news by receiving and processing the constant update flow. As a part of this tracking routine you can set it to strike an alarm if given events occur, like the price of a given stock falling below a critical level. It can also be instructed to make an instant call to your mediaphone (which might be embedded in your wristwatch) when certain events occur. You can then read the critical information on the display.

- Your television becomes more than a television. It receives interactive content of all sorts – 24 h a day.

- When flat screens become mainstream you might have constant update of broadcast news, entertainment and information on the lid of your refrigerator, in your bathroom, or…well, wherever you want.

- Business executives will be able to monitor relevant market conditions constantly in their office, their homes or their summerhouses. More time with the family with still the ability to stay in charge at work.

The second issue to consider is *what content* will be created. We have already gone through a number of examples, but let's just look a bit further into the future:

- Data broadcasting will make it possible to transmit three-dimensional video by surrounding the filmed object with cameras. The viewer can move around in space while the video is viewed. This could, for example, be set up to a live rock concert allowing you to guide yourself around the stage.

- Another possibility is object oriented video. This is video where different parts are active, allowing you to click on sections of the video to dig deeper into the content. Or your PC or TV could know your general product interest and insert background ads on the video that are particularly interesting for you.

- Business and consumer software is no longer sold as versions, but as subscriptions. Upgrades ranging from smaller patches and bug-fixes to major new releases will automatically be broadcast to you as soon as they are released.

- One area having immense possibility is of course computer games. You can imagine live sports games where you sail virtual boats, drive virtual motorcycles or shoot on virtual shooting galleries – in live competition with real sports people. Or you make virtual stock market trades against the real market, fly your virtual aeroplane through a real hurricane (it exists in the real world and is digitally simulated in your game), or play drums and organ on your keyboard to improve the sound of a real live concert – and the result is recorded for replay.

- The large bandwidth in broadcast networks will make it possible to download 3D house models to millions of people. Some may be submitted by viewers, others may be models of real houses. Customers can then move around within the models, and also change the models creating their own dream houses. Or children can submit funny scenarios, build in 3D virtual LEGO, which any receiver can then modify.

Last, but not least, it is likely that data broadcast will introduce interactive services to a large number of people who did not have them before. This is because it can use the "always-on" principle known from radio and television, it does not require a dial-up, it can provide much more compelling media, it requires less battery use on a mobile device, and it can reach people who do not have a modem connection. Some of the possible implications:

- Interactive content can be broadcast via satellite to places that are too poor or remote to have any other on-line connection. It may, for instance, provide distance learning to schools in central Africa and computer games to the children of Australian farmers.

- It may become the natural development of television, as the future television will simply provide this as enhanced functionality.

- It may be the key factor that convince marginal Internet users to get connected.

- It can provide corporate communication solutions.

- It can provide interactive distance learning to people who would otherwise not have a choice.

Data broadcasting is a new challenge and opportunity in the digital economy. David Ticoll, the President of the Alliance for Converging Technologies, once said:

> *"The upheavals of the early 1990s – process re-engineering, downsizing, and PC proliferation – were a tea party compared with what is coming next. Capitalism is about to be completely reinvented. As the ice age of the old economy comes to an end, cracks widen in the fault lines of crumbling business models. Intuition and creativity will blossom in organisations rooted in new media."*

Data broadcasting is the bridging technology that converge broadcasting and the Internet into a single, seamless digital medium. What will follow will be the "intuition and creativity" that David Ticoll spoke about.

ENJOY!

12. Glossary of Terms

ADSL	Asymmetric Digital Subscriber Line (ADSL) is a new modem technology that converts existing twisted-pair telephone lines into access paths for high-speed communications of various sorts.
Asynchronous	A form of data transmission, in which information is sent one character at a time, unrelated to specific timing with regard to transmission.
AVS	Audio Video Streaming (AVS) is a sequence of "moving images" and sound that are sent in compressed form over the medium and displayed by the viewer as they arrive. With audio video streaming, a user does not have to wait to download a large file before seeing the video or hearing the sound. Instead, the media is sent in a continuous stream and is played as it arrives. The user needs a player, which is a special program that uncompresses and sends video data to the display and audio data to speakers. A player can be either an integral part of a browser or distributed by the software manufacturer. Examples of streaming content are news wires, stock tickers, CEO speeches and live concerts.
Bandwidth	The transmission capacity/rate of an electronic line, such as a communications network. The rate can be measured in bits per second, bytes per second, or in cycles per second (Hertz).
Broadband	High-speed data transmission, sufficient to carry multimedia content, such as live video. Commonly used in reference to communication lines at T1 rates (1,544 Mbps) or above; or, defined as any speed above ISDN quality (128 kbps).
Broadcast	A delivery system where a copy of the transmission [does not need to be data only] is given simultaneously to all receivers on a network.
Broadcast channel	Data delivery channels that enable transmission from a sender to multiple receivers.
Broadcast data	Any data that is simultaneously transmitted to multiple recipients on a network.

Broadcast guide	The user will be able to obtain information about the currently available streaming services, package delivery and cached content delivery. The Broadcast guide may present extended programme information (e.g. topic, rating, description, alternative times, etc.). It may also display service information for an extended period of time (e.g. next day, next week, next month, etc.). This basic infrastructure provides interfaces to insert services information from third parties.
Broadcast object	Broadcast objects are mini-channels with server side automation and client side interface. They can be added to other channels to provide a more complete service.
Broadcast request	An element in the schedule representing a file or directory to be broadcast at a specific time.
Broadcaster	The delivery system or person who broadcasts data.
Browser	A software application with a graphical user interface used to view specially coded multimedia data; mostly commonly used to view HTML, XML or SGML coded data.
Cache	An intermediate, high-speed memory on a computer set aside for frequently accessed data to shorten response time.
Cached content delivery	Content packages are delivered and stored in a cache on the user's hard disk or close to the end-user. The cache is automatically managed by the system. Access to the cache is provided through an Internet browser connected to a local HTTP proxy server. This category supports re-purposed and new Internet content.
Channel	An addressable content stream used by the Community Administrator to associate the broadcast of content and/ or applications to end-user subscribers. Channels are administered by the Community Administrator according to customer specific needs.
Client	A computer or process that requests a service from another networked computer system (server).
Client ID	Unique identification number used to identify a subscriber.
Client–server model	The distributed processing paradigm of network systems in which transaction responsibilities are divided into two parts: client and server. Both terms (client and server) can be applied to software programs and also actual computing devices.
Component	An element of a product module.

Component application	A complete, self-contained program that performs a specific function directly for the normal user.
Component process	A complete, self-contained program that performs a specific function directly for the privileged user (i.e. system administrator).
Conditional access	A mechanism to restrict content access to authorised receivers. DVB has recommended a common scrambling algorithm and implementation to provide basic security functions.
Content Provider (CP)	A person or organisation which authors or aggregates content distributed through the platform, along with broadcast guide information.
Customer	The subscriber of the broadcast services.
Customer view	The portion of the SMS window that displays customer information.
Cut-off rate	The maximum time acceptable between programme scheduling is modified and the content/programme is transmitted.
Database administrator	The user who manages the server at the content aggregation side.
DAB	Digital Audio Broadcast
Digital terrestrial	A wireless communication line where the digital signal is transmitted near the ground from one antenna to another using the air as transport media.
Domain	The upper level in the customer organisation, also known as a community.
Download	The transfer of data from a higher level computer to a lower level computer in a hierarchical network context.
DSL/xDSL	Provide high-speed data transfer over existing telephone lines. A reference to all types of digital subscriber lines comprising data rates from 128 kbps to 52 Mbps.
DVB	An international digital broadcast standard for TV, audio and data.
Edit	To make any change of content in a document or file.
Embed	Inclusion of objects like pictures, tables or software units in a HTML document.
Enable/disable	Change locally by a customer, in the state of a channel to be visible or invisible to a customer. Inherent function of the Subscription Management System.
Feature	The beneficial properties of a product or service.
Free-to-air services	The user is able to receive the content free.
Frequency	Repeated occurrences within a given time frame.
Host	Any computer that acts as a server.

HMP	Home Multimedia Platform
Host computer	The primary or controlling computer in a network that provides services such as computation, database access, or special programs.
Host name	The identifying name of any computer that is or can be connected to the Internet.
Hyperlink	A hyperlink is a text item or image in an information item that links the user to another information item in a cross-referencing manner. (Synonymous with hyper-reference.)
Hypermedia	Hypermedia is similar to hypertext, although bringing together many other media, typically of an audio and visual nature.
Hypertext	Hypertext is any text that cross-references other textual information with hyperlinks.
I-Cache	A component on a browser, which acts as a local proxy server, receiving broadcast files then caching them on the local device.
Internet	A collection of networks interconnected by a set of routers that enable them to function as a single, large virtual network.
Intranet	A company's internal network that uses TCP/IP protocol for file transfer, browsing, and communications.
Internet Protocol (IP)	The IP part of the TCP/IP or UDP/IP communications protocol. IP implements the network layer (layer 3) of the protocol, which contains a network address and is used to route data to a network or sub-network. TCP/IP and UDP/IP implement the transport layer (layer 4) of the protocol.
IP multicast	The method of transmitting data to a group of selected users simultaneously as one data stream on an IP network.
Kiosk application	An information system that allows users to, for example, ask questions and order services through the use of a multimedia terminal.
Local area network	A communications network that serves users within a confined geographical area. It is made up of servers, workstations, a network operating system and a communications link.
Locking	In a relational database, a traditional way to guarantee read consistency by giving one process exclusive access to data and making other processes queue up and wait for one-after-another access to the database.

Login/log in	To supply a unique user identification and password to gain access to a system or desktop session.
Logout/log out	To terminate or end access to a system or desktop session.
Media aggregators	Local Area Network) A communications network that serves users within a confined geographical area. It is made up of servers, workstations, a network operating system and a communications link.
Media module	A media module is third party content that fits into the client interface of a channel, but which is either static or updated in a very slow cycle. Examples are the marketing CD-ROM or a distance learning language lesson that is sold as a plug-in to a channel. It can be static or slowly updated, but it is always temporary and does not extend in time.
Modem	Modulator-demodulator. A device that adapts a terminal or computer to an analogue telephone line by converting digital pulses to audio frequencies and vice versa.
Multicast	A packet with a special destination address which multiple nodes on the network may be willing to receive.
Multitasking	A mode of operation offered by an operating system in which the computer works on more than one task at a time, or more than one user can access the same application at the same time.
Narrowband	A slow data transfer rate, usually below 64 kbps. (The transfer rate of an analogue modem is typically 33–56 kbps).
Netscape™ Navigator	Internet browser marketed by Netscape.
Network operator	The network operator runs the broadband network.
Package delivery	Content packages are delivered to the user and stored on his local hard disk. In contrast to cached content delivery, the user can influence how content is handled on reception. (Storage location, how long the content will be kept before deletion.)
Point-to-multipoint	To transmit data from a single source to several recipients simultaneously, also referred to as broadcast.
Polling	The client is instructed to search the server at certain intervals for updated information. Creates illusion of true push by storing data locally in an I-cache on the hard drive.
Port number	Numbers used by TCP/IP to identify the protocol and the end points of communication.

Privilege	The level of permission a user has to read, write, or execute a process.
Process	A process is a set of inter-related resources and activities which transform inputs into outputs, e.g. solution development process, sales process.
Product	A product can is a market offer, consisting of software, hardware, documentation and/or service, consulting including tools that are directly related to the product.
Profiler	A component application on the client that reads the current state of subscription for a client and allows the customer to enable or disable subscribed channels and services.
Protocol	A set of rules defining how networked computers exchange information.
Proxy server	A server (hardware and software or software alone) that is capable of storing web content locally. Requests for the URL based information that is stored in the proxy server can be serviced or provided locally. This improves user response time and reduces network traffic onto the Internet.
Push	In client/server applications, to send data to a client without the client requesting it. The World Wide Web is based on a *pull* technology where the client browser must request a web page before it is sent. Broadcast media, on the other hand, are push technologies because they send information out regardless of whether anyone is tuned in.
Queue	A line or list formed by items in a system waiting for service, following a first-in, first-out constraint.
Router	A device that forwards data packets from one local area network (LAN) or wide area network (WAN) to another.
Schedule database	A database containing schedules of files to be broadcast.
Scheduler™	The Scheduler is a software program that enables content to be broadcast at specified times. It can be automated or performed manually.
Server	A computer or process that provides a service or data to another networked computer system (client).
Service	A more focused multimedia content within a given channel.
Set top box	A box that connects to the television set that receives digital television streams via satellite, cable or other means. A set top box may act as an interface to interactive broadband services.

SMS database explorer
An SMS component application which allows a user to maintain customer, channel, service, and subscription information.

Streaming audio/video
Streaming audio and video content is defined by the possibility of accessing it at any time during a running transmission. The consistency of the content does not depend on previously transmitted data. It is possible to synchronise on an ongoing transmission. Streaming audio and video content is typically not stored on the user's system. Examples of this content category are streaming video (similar to video on MBONE) and streaming audio (Internet radio).

Streaming data
See AVS.

Sub-channel
See Service.

Subscribe/ unsubscribe
Changing, at the remote SMS server, the state of a service to be deliverable or undeliverable to a customer.

Subscription
The state of the relationship between a channel or service and a customer.

Subscription services
The user must register in order to receive the service and there is a fee for the service. This service may require a return path connection.

System administrator
The person who administers and maintains computer systems.

Transmission
Transfer of data from one point to one or many points.

Transparent
In computer use, an adjective describing a device, function, or part of a program that works so smoothly and easily that it is invisible to the user.

Transport
The means by which an object is passed from one process to another.

Unicast
An address which only one receiver will recognise.

Uplink
Portion of a satellite transmission system that makes data travel from a transmitter on the ground to the satellite.

Upload
The transfer of data from a lower level computer to a higher level computer in a hierarchical network context.

ATM
Asynchronous Transfer Mode. A new and faster way of transferring data on a network

BGG
Broadcast Guide Generator. A server module that quickly generates simple electronic program guides (URL-linked program guides) in HTML format.

CMD
Content Meta Data. Information on file-size and status, etc.

DBMS
Database Management System. A software system facilitating the creation and maintenance of a database and the execution of programs using the database.

ETTA	Earliest Time to Accept. The earliest time (the current time plus a set interval) by which a scheduled broadcast can be accepted.
FBS	A server module that allows the user to set in advance when and for what duration one or more files will be broadcast to the client.
FBSR	File Broadcast System Report. The FBS component application that allows user queries to the remote schedule database and downloads the results into a report file.
FSW	The FBS component process which maintains consistency between the local file system and the schedule information in the database.
FTP	File Transfer Process. An FBS component application that broadcasts files. (As opposed to the commonly accepted definition: the Internet protocol and program used to transfer files between hosts.)
GIF	The most important graphics format for the Web.
GUI	Provides the user with a method of interacting with the computer and its special applications, usually via a mouse or other selection device.
GSM	Global System for Mobile Communications. A digital cellular phone technology based on TDMA that is the predominant system in Europe, but is also used around the world.
HTTP	Hypertext Transfer Protocol. An Internet protocol (based on TCP/IP) used to fetch hypertext objects from remote hosts or from the cache of a browser.
IP Address	Internet Protocol Address. A 32-bit number that identifies each sender or receiver of information that is sent in packets across the Internet.
JPEG	Joint Photographic Experts Group. A picture format. Standard for compression of pictures.
LAN	Local Area Network. A corporate/institutional internal client server network.
MAC Address	Media Access Control Address. A hardware device's unique number used to identify its location on a network.
MOTS	Media Object Tracking System. Software for workflow management of multimedia data files.
MPEG	Moving Pictures Expert Group. Established standard for compression of moving pictures and sound.

OSI	Open System Interconnection. The umbrella term for the seven-layer, network architecture model and a series of non-proprietary protocols and specifications developed by ISO as a framework for international standards in heterogeneous computer network architecture.
RDBU	Remote Database Upload. An FBS component process responsible for transferring the schedule database to the network-operating centre.
RFU	Remote File Upload. The FBS component process responsible for transferring files from the local file system to the network-operating centre.
RS	Report Server. The FBS component process responsible for delivering formatted reports on broadcast operations to FBSR users.
SMS	Subscription Management System. A module which manages customer information and subscriptions.
TCP/IP	Transmission Control Protocol/Internet Protocol. The primary communications protocol used over the Internet.
URL	Uniform Resource Locator. The unique location address of an Internet resource composed of protocol and path information.
URN	Uniform Resource Name. A permanent, globally unique name for Internet resources composed of server information. The user can request information with a global name without the specific path information (URL).
VRML	Virtual Reality Mark-up Language. Programming language for three-dimensional worlds.
WAN	Wide Area Network. Geographically dispersed network.
WFIB	Work Flow Information Bus. The WFIB is responsible for transporting content references to and from editors, graphic designers and programmers.

Index

R

radio 3, 83
radio transmission, two types 77
RAND Corporation 15, 27
real time content 192
real-time data streaming *see* data streaming
Real-Time Transport Protocol (RTP) 46
receivers 36–40
reference groups 181
release 202
reliability, definition 100
reliable file transfer *see* package delivery
reporting 134–5, 144–5
requests for development (RFD) 212
research methods 184, 188
retain list, cache management 89
Reuters 224
RFD handling 212
roaming 143
Routing Internet Protocol (RIP) 50
RTP *see* Real-Time Transport Protocol

S

sales cycle 225, 226
Sarnoff, David 42
satellite based channels, redistribution 10
satellite-cable growth cycle 11
satellites 132, 244
    development 4–9
    e-commerce 237
    lifetime expectation 209
    services 8
scalability, data broadcast advantages 154–5
scatter market 229
scheduling
    audio visual streaming 110–11
    booking management 137–43
    cached content 93–4
    data streaming 115–17
    multi-channel 203–5
    non-constant data rate 116
    scheduling tools 140
    transmission chains 142
security 136–7

U
UDP *see* User Datagram Protocol
unicast
   definition 50
   problem with 51
unique media benefits 154–6
upgrades, software 212
usage tracking 96–7
User Datagram Protocol (UDP) 47
user subscriptions 121
user testing 199–201
user-friendly navigation 118–21

V
value chain 36–41, 135, 219–21
   organization process 36
value creation 218–22
value drivers, mature market 240–1
value-added services 222
   provision 41
van Dam, Andy 27
Vertical Blanking Interval (VBI) 55
Very Small Aperture Terminal (VSAT) 70
video, three dimension forecast 244, 247
video-clips, streaming vs file transfer option 205
viewer response 233–5
virtual sport 244
viruses 29, 30
vision, definition 182
VSAT *see* Very Small Aperture Terminal

W
'walled garden', interactive example 34
walled garden layer 124
wallet based approach 98
WAN *see* Wide Area Network
watermarking 148–9
weather channel 193
weather data 112
web based content 82, 93
web-crawlers, definition 94
Whitehouse, Clay 8
Wide Area Network (WAN) 54–5
Winsock 150

'wireless cable' 60
wireless LANs, standards comparison 78–9
Wireless World 4
wireline infrastructure 58–9
workflow management 145–8
WWW (world wide web) origins 27, 28

X
xDSL *see* Digital Subscriber Line
XML *see* eXtensible Mark-up Language